PRACTICAL PROBLEMS in MATHEMATICS

for CARPENTERS

Sixth Edition

Delmar's *PRACTICAL PROBLEMS in MATHEMATICS* Series

- *Practical Problems in Mathematics for Automotive Technicians, 4e*
 George Moore
 Order # 0-8273-4622-0

- *Practical Problems in Mathematics for Carpenters, 5e*
 Harry C. Huth
 Order # 0-8273-4579-8

- *Practical Problems in Mathematics for Drafting and CAD, 2e*
 John C. Larkin
 Order # 0-8273-1670-4

- *Practical Problems in Mathematics for Electricians, 5e*
 Herman and Garrard
 Order # 0-8273-6708-2

- *Practical Problems in Mathematics for Electronics Technicians, 3e*
 Herman and Sullivan
 Order # 0-8273-6761-9

- *Practical Problems in Mathematics for Graphic Communications, 2e*
 Ervin A. Dennis
 Order # 0-8273-2946-3

- *Practical Problems in Mathematics for Health Occupations*
 Louise M. Simmers
 Order # 0-8273-6771-6

- *Practical Problems in Mathematics for Heating and Cooling Technicians, 3e*
 Russell B. DeVore
 Order # 0-8273-7948-X

- *Practical Problems in Mathematics for Industrial Technology*
 Donna Boatwright
 Order # 0-8273-6974-3

- *Practical Problems in Mathematics for Manufacturing, 4e*
 Dennis D. Davis
 Order # 0-8273-6710-4

- *Practical Problems in Mathematics for Masons, 2e*
 John E. Ball
 Order # 0-8273-1283-0

- *Practical Problems in Mathematics for Welders, 4e*
 Schell and Matlock
 Order # 0-8273-6706-6

Online Services

Delmar Online
To access a wide variety of Delmar products and services on the World Wide Web, point your browser to:
http://www.delmar.com/delmar.html
or email: info@delmar.com

thomson.com
To access International Thomson Publishing's home site for information on more than 34 publishers and 20,000 products, point your browser to:
http://www.thomson.com
or email: findit@kiosk.thomson.com

A service of I(T)P®

PRACTICAL PROBLEMS in MATHEMATICS
for CARPENTERS
Sixth Edition

Harry C. Huth

Delmar Publishers Inc.

I(T)P An International Thomson Publishing Company

Albany • Bonn • Boston • Cincinnati • Detroit • London • Madrid • Melbourne
Mexico City • New York • Pacific Grove • Paris • San Francisco • Singapore • Tokyo
Toronto • Washington

NOTICE TO THE READER

Cover Design: Dartmouth Publishing, Inc.

Delmar Staff
Publisher: Robert D. Lynch
Editor: Mary Clyne
Production Manager: Larry Main
Art and Design Coordinator: Nicole Reamer

COPYRIGHT © 1996
by Delmar Publishers
A division of International Thomson Publishing Inc.
The ITP logo is a trademark under license.

Printed in the United States of America

For more information contact:
Delmar Publishers
3 Columbia Circle, Box 15015
Albany, New York 12212-5015

International Thomson Editores
Campos Eliseos 385, Piso 7
Col Polanco
11560 Mexico D F Mexico

International Thomson Publishing - Europe
Berkshire House 168-173
High Holborn
London, WCIV 7AA
England

International Thomson Publishing GmbH
Königswinterer Strasse 418
53227 Bonn
Germany

Thomas Nelson Australia
102 Dodds Street
South Melbourne, 3205
Victoria, Australia

International Thomson Publishing - Asia
221 Henderson Road
#05-10 Henderson Building
Singapore 0315

Nelson Canada International
1120 Birchmount Road
Scarborough, Ontario
Canada, M1K 5G4

International Thomson Publishing - Japan
Hirakawacho Kyowa Building, 3F
2-2-1 Hirakawacho
Chiyoda-ku, Tokyo 102
Japan

10 9 8 7 6 5 4 xx 99 98

Library of Congress Cataloging-in-Publication Data
Huth, Harry C.
 Practical problems in mathematics for carpenters / Harry C. Huth.
 — 6th ed.
 p. cm. — (Delmar's practical problems in mathematics series)
 ISBN 0-8273-6987-5
 1. Carpentry—Mathematics. 2.Carpentry—Problems, exercises,
etc. I. Title II. Series.
TH5612.H87 1995 95-9414
513'.024694—dc20 CIP

Contents

Preface

Practical Problems in Mathematics for Carpenters, Sixth Edition, has been written to give students who are interested in carpentry practice in solving problems common to the carpentry trade. The student's comprehension of terminology used in the field of carpentry is strengthened through the use of this workbook.

Practical Problems in Mathematics for Carpenters is one in a series of books designed to offer students practical problem-solving experience in various occupations. The *Practical Problems in Mathematics* series offers a step-by-step approach to the mastery of basic mathematics skills. Each book includes relevant and easily understood problems in a specific vocational field. A review of denominate numbers, a glossary, and answers to odd-numbered problems are also included at the end of the workbook.

Revisions in the Sixth Edition have increased the percentage of problems devoted to realistic carpentry applications. Costs of materials and labor, as well as the types of materials used, have been updated for realism. Girders, sills, and floor joists have been combined into a single unit without omitting any carpentry techniques.

The *Instructor's Guide* provides worked-out solutions to most of the problems and answers to all the problems found in the workbook. A summary of all formulas used in the workbook is included for convenience.

USING A CALCULATOR

Today, the availability of low-cost calculators has made doing mathematics calculations much easier. A complicated mathematics problem can be solved more quickly and with less effort.

DISADVANTAGES

Some pitfalls do exist, and we *must* be aware of them before using a calculator. These pitfalls can usually be avoided *if an estimate is made in advance* of turning on the calculator. When using the calculator, always compare its answer with your estimate. If they do not agree, repeat the solution process very carefully. You might have incorrectly placed, or even omitted, a decimal point.

ADVANTAGES

1. Calculators are a great help in changing fractions to decimals. However, there is no reverse gear. You cannot change decimals to fractions.

2. When working with percentages, a calculator can change a percentage to a decimal, for example, 42.6% to .426. This time-saver also helps you avoid possible errors.

3. When doing a series of operations on a calculator, you do not have to get an answer at each step. Only after going through all the steps do you press the "equals" button.

4. A hidden advantage of estimating, then using a calculator, is that your skill improves as a result of continuously estimating.

When purchasing a calculator for use in carpentry, there is no need to get an expensive one with more functions than you will be using on a regular basis. Addition, subtraction, multiplication, division, percent, and possibly a square root capability will be all that you need. This type is available in inexpensive solar models with which there is no concern about battery life and infrequent use.

An instruction booklet comes with most calculators. Read it completely!

Whole Numbers

 Unit 1 ADDITION OF WHOLE NUMBERS

PRACTICAL PROBLEMS

Add the following quantities:

1.
```
     13 feet
     16 feet
      8 feet
     24 feet
   + 51 feet
```

4.
```
     270 pounds
     140 pounds
     368 pounds
     609 pounds
   + 306 pounds
```

7.
```
       261 square inches
     1,094 square inches
       861 square inches
       644 square inches
   + 1,326 square inches
```

2.
```
   1,312 meters
     644 meters
      31 meters
   3,609 meters
   +  234 meters
```

5.
```
   $1,608
      914
       83
   +1,144
```

8.
```
      23 pounds
   1,116 pounds
   2,492 pounds
   + 3,844 pounds
```

3.
```
     3,002
       994
       681
   + 2,411
```

6.
```
   $12,492
     2,063
     3,876
   + 9,432
```

9.
```
   15,642 board feet
    8,431 board feet
   13,092 board feet
   + 4,467 board feet
```

10. How many linear feet of cove molding are needed to trim the ceiling of a 12-foot by 16-foot family room? _____

11. What is the total length of the house shown on the partial floor plan? _____

12. A basement has a recreation room, store room, and furnace room. The floor spaces are 286 square feet, 91 square feet, and 164 square feet, respectively. What is the total number of square feet in this floor space? _____

13. Four houses require 1,800 square feet, 1,050 square feet, 1,200 square feet, and 1,460 square feet of asphalt shingles, respectively. How many square feet of shingles are needed for the four houses? _____

14. A crew of four worked 24 hours, 18 hours, 20 hours, and 22 hours. What is the total number of hours worked? _____

15. A contractor paid bills of $2,480, $765, $1,446, $1,011, and $1,610 for materials. Find the total cost of the materials. _____

16. In making five excavations, the following cubic meters of earth were removed: 5,040; 6,070; 7,940; 11,424; and 35,216. Find the total number of cubic meters of earth removed. _____

17. Three deliveries of ½-inch by 6-inch siding are as follows: 2,450 board feet, 2,760 board feet, and 2,875 board feet. What is the total number of board feet delivered? _____

18. A carpenter lays 1,300 wood shingles the first day, 1,400 the second, and 1,500 the third. How many shingles are used in three days? _____

19. What is the total number of square feet of floor underlayment needed to complete a job if three rooms require 120 square feet, 90 square feet, and 300 square feet, respectively? _____

20. A contractor buys 14,500 board feet of 1-inch white pine, 1,250 board feet of 2-inch spruce, and 1,450 board feet of 3-inch hemlock. Find the total number of board feet.

21. The cost of material for a remodeling job is as follows: lumber, $476; masonry, $148; hardware, $62; and paint, $85. What is the total cost of materials?

22. In estimating the finish flooring for a house, a contractor lists the room areas as follows: living room, 168 square feet; dining room, 152 square feet; bedroom, 142 square feet; hall, 45 square feet; and kitchen, 125 square feet. What is the total area to be floored?

23. The following items of framing lumber are ordered for a housing development: 1,472 pieces of 2-inch by 4-inch studs; 1,627 pieces of 2-inch by 10-inch joists; 1,827 pieces of 2-inch by 6-inch studs; and 572 pieces of 2-inch by 8-inch joists. How many pieces of framing lumber are ordered?

24. A residence measures 28 feet by 54 feet. Plans show a 2-inch by 6-inch single sill plate. How many linear feet of sill plate are needed?

25. The amounts of materials needed to construct the illustrated outbuilding are shown in the chart. Determine the total number of board feet required for the job.

	Material	Quantity
A	Sill	64 board feet
B	Plate stock	148 board feet
C	Studding	380 board feet
D	Collar beams	48 board feet
E	Rafters	268 board feet
F	Headers	40 board feet
G	Siding	960 board feet
	Total	? board feet

26. A contractor is building four buildings. This is the order of supplies for the concrete block foundations. Find the amount of each material needed for the total order.

Foundation Sizes	a. Concrete Blocks	b. Masonary Cement	c. Sand
A 50' × 32' × 8'	1,476 blocks	49 bags	7 tons
B 24' × 30' × 8'	971 blocks	32 bags	5 tons
C 52' × 36' × 10'	1,979 blocks	66 bags	9 tons
D 20' × 12' × 9'	647 blocks	23 bags	3 tons

Totals a. _____ b. _____ c. _____

Note: The distance around a building is called its *perimeter*. This measurement is often required by carpenters, masons, and other tradespeople in order to estimate quantities of materials needed. To find the perimeter of a building:

a. For rectangular-shaped buildings, add the length and width together and double the result; or

b. Add all the sides together.

27. Find the perimeters of the following rectangular-shaped buildings:

a. 30'–0" × 45'–0" a. _____

b. 52'–0" × 172'–0" b. _____

c. 36'–0" × 102'–0" c. _____

d. 49'–0" × 116'–0" d. _____

e. 62'–0" × 214'–0" e. _____

f. 48'–0" × 84'–0" f. _____

g. 29'–0" × 84'–0" g. _____

h. 46'–0" × 49'–0" h. _____

i. 51'–0" × 215'–0" i. _____

j. 170'–0" × 60'–0" j. _____

k. 35'–0" × 115'–0" k. _____

l. 101'–0" × 45'–0" l. _____

Note: To construct concrete sidewalks, garage floors, and driveways, it is often necessary to use a piece of 2-inch by 4-inch lumber (or some other suitable size), placed on edge, as a form to hold the concrete until it is set.

28. A garage floor and driveway are poured in a single slab. The solid lines in the diagram indicate location of forms. Find the number of linear feet of 2 × 4s needed to allow the forms for the driveway and garage as shown. (Do not make allowance for joints at the corners of the forms.) _____

29. The outside forms for the footings of this house are one board high. Calculate the number of linear feet of 2-inch by 8-inch stock needed for the job. (Do not make allowance for joints.) _____

Unit 2 SUBTRACTION OF WHOLE NUMBERS

PRACTICAL PROBLEMS

Subtract the following quantities:

1. 120 inches
 − 43 inches

2. 1,473 pints
 − 611 pints

3. 909 ounces
 − 640 ounces

4. 1,636
 − 703

5. 3,469 square feet
 − 983 square feet

6. 2,091 yards
 − 993 yards

7. 12,643 gallons
 − 7,123 gallons

8. 16,298 kilograms
 − 9,696 kilograms

9. 23,122
 − 10,069

10. $14,254
 − 9,676

11. 55,864 board feet
 − 34,964 board feet

12. $25,150
 − 11,909

Note: No allowance is made for the saw kerf.

13. A length of stock 106 inches long is cut from a board 120 inches long. What is the length of the remaining piece? _____

14. The supply of plywood in stock is 9,984 square feet. After 6,976 square feet are used, how much is left? _____

15. From a supply of 7,326 linear feet of baseboard, a total of 4,560 feet is used. How many feet of baseboard remain in the supply? _____

16. The area of a shop floor is 1,650 square feet. How much area remains to be painted after 735 square feet are covered? _____

17. The excavator for the foundation of a house must remove 650 cubic meters of earth. How much remains to be excavated after 175 cubic meters are removed? _____

18. A lumber dealer has 632,000 board feet of white pine on hand. After the sale of 328,582 board feet, how many board feet remain? _____

19. A carpenter builds a redwood deck for $1,450. The material, labor, and other costs total $735. What is the profit? _____

20. A contractor buys 6,000 board feet of oak flooring. One house requires 1,928 board feet and another needs 1,850 board feet. How much flooring is left? _____

21. The balance in a contractor's checking account is $1,176. If $321 is withdrawn, what is the new balance? _____

22. Shown below is the floor plan of a two-car detached garage. What is the width of the wall space A at the front of the garage? _____

23. What is the distance from the outside of the rear wall of the garage to the center of the side door opening, dimension B? _____

24. What is the length of missing dimension C at the rear of the garage? _____

25. Determine missing dimension A in the sketch. _____

26. Determine missing dimension B in the sketch. _____

27. Find missing dimension X in the sketch. _____

28. Find missing dimension D in the sketch. _____

29. Find the length of missing dimension E in the sketch. _____

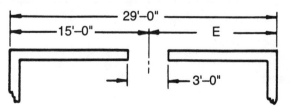

30. A piece of lumber is 18 feet long. After a piece 4 feet long is sawed off, what length is left?

31. From an inventory of 1,272 sheets of ¼-inch paneling, a dealer sold the following numbers of sheets: 15, 73, 87, 121, 53, 22, and 30. How many sheets of paneling are left?

Note: When locating the position of a window in a wall during construction, the distance that the opening is placed from one end of the building is usually given on the plans. In the following problems, find the distance from the window to the *other* end of the building.

Note: Use this illustration for problems 32–35.

Prob. No.	Total Wall Length Distance A	Width of Window Distance C	Distance B from End of Building to Window
32.	32'–0"	3'–0"	8'–0"
33.	16'–0"	3'–0"	5'–0"
34.	30'–0"	4'–0"	12'–0"
35.	34'–0"	5'–0"	13'–0"

32. _____

33. _____

34. _____

35. _____

Unit 3 MULTIPLICATION OF WHOLE NUMBERS

PRACTICAL PROBLEMS

Multiply the following quantities:

1. 16 inches
 × 44

2. 56 liters
 × 17

3. 254
 × 16

4. 352
 × 75

5. 1,809 inches
 × 62

6. 2,834 pounds
 × 170

7. 659 yards
 × 212

8. 350 hours
 × 521

9. $1,205
 × 57

10. 1,972 board feet
 × 109

11. 1,257 gallons
 × 857

12. $16,005
 × 77

13. What is the total length of 25 pieces of door casing if each piece is 7 feet long? _____

14. Determine the total wall area of a room if each of the four walls has an area of 357 square feet. (Do not allow for openings.) _____

15. A carpenter places 200 linear feet of joists per hour. How many feet are placed in 8 hours? _____

16. The cost of a new garage is $9,755. Find the total cost of building 13 garages of this kind. _____

17. Find the cost of 28 squares of asphalt shingles at $37 per square. (A square is 100 square feet.) _____

18. Allowing 760 shingles per square (100 square feet), how many shingles are required to cover 24 squares? _____

19. The illustration shows the dimensions of a gable roof that is to be shingled. Determine the number of square feet that the entire roof contains. (Assume that the opposite side has no projection.) _____

20. The area of a rectangle is found by multiplying the length by the height. A partition is 24 feet long by 8 feet high. Find the area of one side of the partition. _____

21. If there are 250 cedar shingles in one bundle, how many are there in 258 bundles? _____

22. In making a table, 19 board feet of oak are used, including waste. How much stock is needed to complete an order for 54 such tables? _____

23. A breakfast set requires 43 board feet of lumber. How many board feet are needed for 53 sets that are to be made for a coffee shop? _____

24. Three hundred twenty galvanized nails are required to apply a square of shingles. How many nails are needed to lay 25 squares? _____

25. How many linear feet of furring are there in 9 bundles, each containing 10 pieces that are 12 feet long? _____

26. How many board feet of lumber are there in the following list of materials: 164 joists—32 board feet each; 16 girders—96 board feet each; 58 rafters—28 board feet each; 14 posts—44 board feet each; 296 boards—16 board feet each; 164 studs—8 board feet each. _____

Note: To solve more complex problems, try to divide the problem into a series of easier steps. Do problems 27 and 28 by following the steps outlined and using the illustrations given.

27. How many square feet of floor space are there in the building shown?

a. Find missing dimension A and B.

A = _____

B = _____

b. Divide the total area into rectangular Sections I and II. (Other correct divisions are possible.) Find the areas of Sections I and II.

(*Area = length × width*)

Section I = _____

Section II = _____

c. Add the areas of Section I and II to find the total area.

Total = _____

28. The walls of the building are 11 feet high. Find the total area of the outside walls.

a. Write dimensions A and B, found in problem 27.

A = _____

B = _____

b. Find the area of each wall.

(*Area = length × height*)

u = _____

v = _____

w = _____

x = _____

y = _____

z = _____

c. Find the total area.

Total = _____

Unit 4 DIVISION OF WHOLE NUMBERS

PRACTICAL PROBLEMS

Divide the following quantities:

1. 846 inches ÷ 6 = _____

2. 8,916 pounds ÷ 6 = _____

3. 15,553 ÷ 2 = _____

4. 6,921 ÷ 14 = _____

5. 3,434 centimeters ÷ 34 = _____

6. $26,334 ÷ 77 = _____

7. 4,068 ÷ 142 = _____

8. 26,104 grams ÷ 26 = _____

9. 217,124 millimeters ÷ 206 = _____

10. How many pieces of lumber 36 inches long can be cut from a piece of lumber 190 inches long? (Do not allow for saw kerfs.) _____

11. How many hours are needed to lay 902 square feet of subflooring at the rate of 82 square feet per hour? _____

12. One person can apply 24 square feet of siding per hour. At the same rate, how long does it take a crew of 6 to put on 11,952 square feet of siding? _____

13. How many hours does it take to install 3,192 square feet of batt-type insulation at the rate of 152 square feet per hour? _____

14. How many joists spaced 16 inches o.c. (on center) are required for a floor 624 inches long? _____

 Note: Add one joist for a starter.

15. Find the number of studs spaced 16 inches o.c. required for a load-carrying partition 768 inches long.

 Note: Add one stud for a starter.

16. A common gable roof is 34 feet long. How many rafters spaced 2 feet o.c. are required for one side of the roof?

 Note: Add one rafter for a starter.

Note: Use this illustration for problems 17 and 18.

17. The illustration shows a short flight of steps. Dimension A is the height or rise of each step. Determine dimension A.

18. Determine the run of each step as indicated by dimension B in the illustration.

19. A gable roof is 72 feet long. Trusses are spaced 2 feet o.c. How many trusses are needed for the roof?

 Note: Add one truss as a starter.

20. The main stairway in a building has 17 risers. If the story height (distance from top of the first-story floor to the top of the second floor) is 119 inches, what is the height of each riser?

Note: For problems 21–25, allow for only one set of joists.

21. How many floor joists, as illustrated, are required for a building 576 inches long if the joists are placed 16 inches o.c.?

22. How many joists, spaced 16 inches o.c., are needed for a building that is 432 inches long? _____

23. How many joists, spaced 16 inches o.c., are required for a building that is 720 inches long? _____

24. The specifications for a building state that the joists are to be placed 2 feet o.c. How many floor joists are required if the building is 116 feet long? _____

25. A warehouse is 116 feet long. To support the heavy load on the floor, joists are spaced 1 foot o.c. How many joists do the specifications require? _____

26. A warehouse is 98 feet wide by 164 feet long. Girders run the long way of the building and are spaced 14 feet o.c. How many linear feet of girder are required? _____

27. A building is 42 feet wide and 80 feet long. Girders running the length of the building are spaced 6 feet o.c. The outside walls rest on a foundation. Supporting each girder are piers spaced 5 feet o.c. How many piers are needed? _____

28. Determine the number of sheets of 4-foot by 8-foot plywood subfloor needed to cover the floor area as shown.

Unit 5 COMBINED OPERATIONS WITH WHOLE NUMBERS

PRACTICAL PROBLEMS

Review and apply the principles of addition, subtraction, multiplication, and division of whole numbers.

1.	629	2.	4,567	3.	642	4.	833
	315		632		− 218		− 394
	783		7,992				
	+ 324		+ 6,415				

5. 7,035
 × 476

6. 2,884
 × 3,526

7. 39) 936

8. 252) 124,992

9. A carpenter cuts three pieces of weather stripping from a roll 120 inches long. How much remains if the pieces were 37 inches, 34 inches, and 36 inches? _____

10. How much baseboard will be required for a room 12 feet by 15 feet if 36 inches are allowed for a door? _____

11. A contractor hires three carpenters to work 9 hours a day for 5 days and two carpenters to work 4 hours for 1 day. What is the total number of hours they worked? _____

12. How many squares of roofing will be needed for a roof that is 25 feet by 60 feet? (One square covers 100 square feet.) _____

13. How many pieces of 4-foot by 8-foot plywood subfloor will be needed for the floor in the diagram? _____

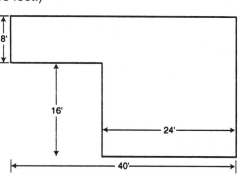

14. Three helpers are hired and paid $5 per hour. One helper works 6 hours, another 8 hours, and the third only 4 hours. What is the total amount the carpenter paid?

15. How many square feet of flooring will be needed for a cottage with only four rooms measuring: 9 feet by 12 feet, 8 feet by 10 feet, 13 feet by 11 feet, and 19 feet by 16 feet?

16. After a carpenter installed baseboards and ceiling molding throughout a house, the owner changed his mind and wanted a different style. The carpenter worked 50 minutes of every hour removing and replacing the trim. It took 2½ hours to remove the trim and 5¼ hours to replace it with new trim. How many hours did he actually work?

17. Floor joists are delivered in 16-foot lengths. If only 14½-foot lengths are needed, how much waste results when 11 joists are used?

18. A 6-floor apartment building has an average of 7 doors per apartment, and there are 5 apartments on each floor. How many doors are required?

19. A porch with eight windows is being fitted with new screens. Six windows measure 30 inches by 4 feet, and the other two, 18 inches by 4 feet. Making no allowance for waste, how many square feet of screening should be ordered?

20. Five pieces of stock, each 120 inches, are cut into 106-inch lengths. Find the total amount of waste.

21. A sawmill has 750,000 board feet of hemlock. After selling 475,000 board feet, the remainder is sold at $110 per 1,000 board feet. What is the amount of the final sale?

22. The cost of building garages at a new development is $10,825 per garage. If a subcontractor builds five and then sells four at $12,500 each to the developer, how much does the subcontractor now have invested?

Common Fractions

Unit 6 ADDITION OF COMMON FRACTIONS

BASIC PRINCIPLES OF ADDITION

A fraction is a part of a whole. A whole inch may be divided into equal parts in many ways. For instance, an inch may be divided into eight equal parts. Each part is known as an eighth of an inch (⅛ inch).

EIGHTHS

Should five of these parts be needed in measuring a length, this quantity would be five-eighths inch and be written ⅝ inch.

The *number of parts* is called the *numerator*. ⟶ $\dfrac{5}{8}$ ⟵ The *whole* is called the *denominator*.

To add fractions, each fraction must first be expressed with a denominator common to all other fractions in the problem.

Example: ½ + ⅓ + ⅝ + ⁴⁄₉ = ¹⁸⁄₃₆ + ¹²⁄₃₆ + ³⁰⁄₃₆ + ¹⁶⁄₃₆

A *common denominator* may always be found by multiplying all the denominators of the problem; this often results in an extremely large number.

Example: ½ + ⅓ + ⅝ + ⁴⁄₉ (2 × 3 × 6 × 9 = 324)

A *least common denominator* (LCD) is generally found by inspection. If inspection is not obvious, it may always be determined as follows:

1. Factor every denominator into its smallest factors.

2. Find the product of all different factors, using each factor the greatest number of times it is found in a single denominator. This product is the LCD.

3. Multiply the original fraction—both numerator and denominator—by the factor of the LCD that is missing in the original denominator.

Example: $\frac{5}{6} + \frac{4}{9}$

$$6 = 2 \times 3$$
$$9 = 3 \times 3$$

The different factors are 2 and 3; 3 appears twice in 9. Therefore, the LCD is 18, or

$$2 \times 3 \times 3$$

Multiply the original fractions and add as follows:

$$\frac{5}{6} = \frac{5}{2 \times 3} = \frac{5}{2 \times 3} \times \frac{3}{3} = \frac{15}{18}$$

$$\frac{4}{9} = \frac{4}{3 \times 3} = \frac{4}{3 \times 3} \times \frac{2}{2} = \frac{8}{18}$$

$$\frac{23}{18} = 1\frac{5}{18}$$

4. Add the numerators and express the sum over the common denominator.

Example: $\frac{1}{2} + \frac{1}{3} + \frac{5}{6} + \frac{4}{9}$

Again, the LCD is 18.

$$\frac{1}{2} = \frac{1}{2} \times \frac{3}{3} \times \frac{3}{3} = \frac{9}{18}$$

$$\frac{1}{3} = \frac{1}{3} \times \frac{2}{2} \times \frac{3}{3} = \frac{6}{18}$$

$$\frac{5}{6} = \frac{5}{2 \times 3} \times \frac{3}{3} = \frac{15}{18}$$

$$\frac{4}{9} = \frac{4}{3 \times 3} \times \frac{2}{2} = \frac{8}{18}$$

$$\frac{38}{18} = 2\frac{1}{9}$$

PRACTICAL PROBLEMS

Add the following fractional quantities:

1. ¾ inch + ⅝ inch _____

2. ⅛ + ⅓ + ⅙ _____

3. ⅜ gallon + ⅞ gallon _____

4. ¼ yard + ½ yard + ⅓ yard _____

5. ⅚ square foot + ¾ square foot + ⅜ square foot + ⅓ square foot _____

6. 3 board feet + 7½ board feet + 19¾ board feet _____

7. ⅔ hour + 2¼ hours + 4⅓ hours _____

8. What is the total thickness of three pieces of plywood ⁵⁄₁₆ inch, ⅝ inch, and
 ⅞ inch thick? _____

9. What is the total thickness of a table top made of ¾-inch particle board
 covered with ¹⁄₁₆-inch laminated plastic? _____

10. A desk top is ¾ inch thick. It is covered with plate glass ¼ inch thick. Find
 the total thickness. _____

11. Find the total thickness of three pieces of plywood ⅝ inch, ⅜ inch, and ¾
 inch thick. _____

Note: Use this illustration for problems 12–14.

12. A section of the outside wall of a
 frame building is shown. What is
 the wall thickness? _____

WALL
THICKNESS

$\frac{1''}{2}$ INSULATION BOARD

$3\frac{1''}{2}$ STUD

13. The insulation board sheathing
 on the exterior face of the wall in
 the illustration is covered with
 patterned cedar shingles. These
 add ⅝ inch to the thickness.
 What is the thickness of the wall
 after it is shingled? _____

14. After applying ⅝-inch cedar
 shingles to the exterior, the
 interior face of the stud is
 covered with drywall ½ inch
 thick. What is the total thickness
 of the wall? _____

15. Drywall partitions separate the rooms of a house. If the partition studs are 3½ inches thick, and the drywall on each face of the stud is ½ inch thick, what is the thickness of the partition?

16. How thick is a panel built up of plywood ⅜ inch, ¼ inch, ⅝ inch, and ¾ inch thick?

17. The base of a platform is built of lumber 2⅝ inches thick. It is covered with boards ¾ inch thick. Find the total thickness.

18. A table top ⅞ inch thick is covered with ¹⁄₃₂-inch veneer. What is the total thickness of five of these table tops?

19. A shed door is made of two pieces of ½-inch and one piece of ⅜-inch plywood. Find the total thickness of the door.

Note: Use this illustration for problems 20–24. All stock is ¾ inch thick. Shelves have ⅜-inch-deep dados.

20. What is the overall height of the illustrated bookcase?

21. What is the overall length of shelf A of this bookcase?

22. How high from the floor is the top of shelf A?

23. Find the length of the top of the bookcase.

24. What is the height of the finished pieces used for sides of the bookcase?

25. Using the entire width of the board, a carpenter cuts three pieces of stock of different widths from one board. The first is 2¼ inches, the second is 3⁵⁄₁₆ inches, and the third is 4¼ inches wide. If ⅛ inch is allowed for each saw kerf, find the width of the board used.

26. The foreman on a job gives an apprentice carpenter the job of ripping a ⅞-inch piece, a 1¼-inch piece, a 2⁷⁄₁₆-inch piece, and a 3¾-inch piece from a 4-foot by 8-foot panel of ¾-inch plywood. If ⅛ inch is allowed for each saw kerf, how much of the panel's width is used?

Note: Use this illustration for problems 27 and 28.

27. What is the vertical distance A?

28. What is the horizontal distance B?

29. How much is taken from the depth of the drawer shown in the illustration if the groove is ⅜ inch from the bottom and the drawer bottom is ¼ inch thick?

30. What is the total width of the cornice shown in the illustration below? _____

31. What is the height of the chest shown below? _____

Unit 7 SUBTRACTION OF COMMON FRACTIONS

BASIC PRINCIPLES OF SUBTRACTION

As in addition, fractions must be expressed with common denominators before subtracting. Subtract the numerators and express the difference over the common denominator.

Example: Subtract ½ from ⅔

$$\frac{2}{3} - \frac{1}{2} = \frac{4}{6} - \frac{3}{6}$$ Express with common denominators

$$\frac{4 - 3}{6} = \frac{1}{6}$$ Subtract numerators

To subtract a fraction from a whole number or a mixed number, express the whole number as an improper fraction, with a denominator common to the denominator of the fraction.

Example: ⅔ from 5

$$5 - \frac{2}{3} = \frac{15}{3} - \frac{2}{3}$$ Express with common denominator

$$\frac{15 - 2}{3} = \frac{13}{3}$$ Subtract numerators

$$\frac{13}{3} = 4\frac{1}{3}$$ Express as a mixed number

PRACTICAL PROBLEMS

1. Subtract ⅝ inch from ²⁹⁄₃₂ inch. _____

2. Subtract ¾ inch from ³¹⁄₃₂ inch. _____

3. Find the thickness of a board ¾ inch thick after ¹⁄₁₆ inch is planed off one side. _____

4. A rough board 5⅞ inch wide has ⅛ inch taken off by planing on one edge. What is its width? _____

5. A countertop is laminated with $\frac{1}{16}$-inch plastic. The total thickness is $\frac{13}{16}$ inch. What is the thickness of the top without the laminate? _____

6. How much must a $\frac{7}{8}$-inch board be planed to make it the required thickness of $\frac{25}{32}$ inch? _____

7. The thickness of a piece of three-ply wood is $\frac{1}{2}$ inch. It was made by gluing $\frac{1}{8}$-inch veneer to both faces of a core. Find the thickness of the core stock. _____

Note: Nails are measured in *penny* or *d* units. A 2 penny nail, abbreviated 2d, is 1 inch long. Lengths of nails are found on a nail chart. Use this chart to solve problem 8.

Penny (d) Size	Length in Inches	Wire Guage			
		Common Nails	Box Nails	Finish Nails	Casing Nails
2	1	15	16	17	16
3	1¼	14	15	16	15
4	1½	12	14	15	14
5	1¾	12	13	14	14
6	2	11	12	13	13
7	2¼	...	12	12	12
8	2½	10	11	12	11
10	3	9	11	11	11
12	3¼	9	...	11	10
16	3½	8	...	11	10
20	4	6	...	10	9

NAIL CHART

8. How much longer is a 12d nail than an 8d nail? _____

9. Find the difference in width of two pieces of hardwood flooring $\frac{25}{32}$ inch by 2⅝ inches and $\frac{25}{32}$ inch by 1⅝ inches _____

Note: Answers for all problems should be in simplest form. For example,
3'–21½" = 3' + 12" + 9½" = 3' + 1' + 9½" = 4'–9½"

10. As shown in the illustration, the rail for a table is to have a ¾-inch tenon cut on each end. If the finished length of the stock is 30 inches, what is the distance between the shoulders?

TENON

Note: Use this illustration for problems 11–15.

11. The illustration shows an oblique view of a platform. The top is ¾ inch thick. How long are the legs?

12. The legs on this platform are 2¼ inches square. What is the length of the rail between the legs?

13. If the platform top is ¾ inch thick, what is the width of the rail?

14. Find measurement B from the outside face of one platform leg to the outside face of the other leg.

15. How long should the rail be cut if 1¼ inches are added on each end for the tenons? (The legs are 2¼ inches square.)

16. Turned spindles for a room divider are purchased in standard lengths of 60 inches. The opening in which they are used has a height of 56⅞ inches. How much must be cut off each spindle?

17. How much should be cut from a piece of molding 6⅞ feet long in order to have a piece 6¼ feet long?

18. What is the final thickness of a 3-inch piece of material after ⅛ inch is planed off *both* faces?

Note: Use this illustration for problems 19–22. All stock is ¾-inch thick. Shelves have dados ⅜-inch deep.

19. What is the overall length of the shelves in this cabinet?

20. What is the inside distance between the sides of this cabinet?

21. The shelves are spaced 11 inches, top-to-top. What is the distance between two shelves?

22. The back of the cabinet is ¾ inch thick. What is the depth of the shelves if the back sets flush with the back edge of the sides?

33¾"

¾"

12⅛"

42"

Note: Use this illustration for problem 23.

23. Each shutter installed in a "pass-thru" opening has a height of 25⅞ inches. How much must be cut off the top of each shutter to make it fit?

39 ½″

25 ½″

Unit 8 MULTIPLICATION OF COMMON FRACTIONS

BASIC PRINCIPLES OF MULTIPLICATION

When multiplying two or more common fractions multiply the numerators and then the denominators. Finally, reduce the fraction to the lowest terms.

Example: $\dfrac{3}{4} \times \dfrac{2}{5} = \dfrac{3 \times 2}{4 \times 5} = \dfrac{6}{20}$ Multiply numerators and denominators

$\dfrac{6}{20} = \dfrac{3}{10}$ Reduce to lowest terms

When multiplying fractions by whole numbers, change the whole number to a fraction. Proceed as in multiplying common fractions.

Example: $\dfrac{5}{8} \times 3 = \dfrac{5}{8} \times \dfrac{3}{1}$ Change to a fraction

$\dfrac{5}{8} \times \dfrac{3}{1} = \dfrac{15}{8}$ Multiply numerators and denominators

$\dfrac{15}{8} = 1\dfrac{7}{8}$ Reduce to lowest terms

When multiplying fractions by mixed numbers, change the mixed number to an improper fraction. Then proceed as in multiplying common fractions.

Example: $\dfrac{1}{3} \times 3\dfrac{3}{5} = \dfrac{1}{3} \times \dfrac{18}{5}$ Change to improper fraction

$\dfrac{1}{3} \times \dfrac{18}{5} = \dfrac{18}{15}$ Multiply numerators and denominators

$\dfrac{18}{15} = 1\dfrac{1}{5}$ Reduce to lowest terms

PRACTICAL PROBLEMS

Multiply the following quantities:

1. ⅝ × ⅜ inch _____

2. ⅞ × ⅘ _____

3. 2⅜ × 4¼ gallons _____

4. 5⅜ × 2⅞ inches _____

5. 10⅓ × 3⅔ _____

8. ¾ × ⅝ _____

6. ⅞ × ⅝ _____

9. ⅞ × ¼ _____

7. ¾ × ⅛ board foot _____

10. ⅔ × ⅞ _____

11. A wall is paneled with 38 boards. Each board has a 3⅜-inch exposed surface. Find the width of the wall in inches. _____

12. There are 14 risers in the stairs from the basement to the first floor of a house. Find the total height if each rise is 7⅛ inches high. _____

13. Shingles are laid so that 4¾ inches are exposed in each course. How many vertical feet of roof are covered by 28 courses? _____

14. A board 5¾ inches wide is cut to ¾ of its original width. find the new width. _____

15. A beam is 25½ feet long. Find ¾ of its length. _____

 Note: In answer, convert fraction of foot to inches by multiplying by 12.

Note: Use this illustration for problems 16 and 17.

16. The flight of stairs shown has 9 risers. The height of each riser is 7⅜ inches. What is the height from floor to floor (total rise)? _____

17. The run of each step on the stairs is 9⅜ inches. What is the total run of the flight? _____

18. A flight of stairs has 14 risers, each 7¼ inches high. What is the total rise? _____

19. A piece of plywood is ½ inch thick. It is increased by gluing two ⅜-inch pieces to it. How thick is the finished product? _____

20. A carpenter can place 9 floor joists in ¼ hour. At this rate, find the time necessary to place 126 floor joists. _____

21. If ¼ inch on a drawing represents 1 foot, how many inches on the drawing represent 18 feet? _____

22. Four pieces of wood each 11⅜ inches long are needed to build a cabinet. To get all four pieces from one board, how long must that board be? Allow 1 inch for waste. _____

23. A board is to be ripped into four strips. Each one must be 2¾ inches wide. Allowing ⅞ inch for waste in milling the stock, what width board should be used? _____

24. Find the number of hours required to install 15 roof trusses if it takes ⅙ hour per truss. _____

Note: Use this illustration for problems 25 and 26.

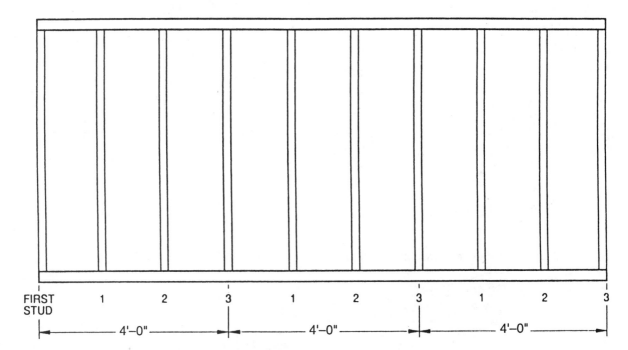

FIRST STUD 1 2 3 1 2 3 1 2 3

|← 4'–0" →|← 4'–0" →|← 4'–0" →|

Note: When studs are placed 16 inches o.c., there are 3 studs for every 4 linear feet of wall, plus one to start the wall. See figure. Estimators calculate the number of studs in a wall by multiplying the length of the wall in feet times ¾ and adding one (the first stud).

25. How many studs are needed for a wall 24 feet long with studs placed 16 inches o.c.? _____

26. How many studs are needed to frame a wall 36 feet long with studs placed 16 inches o.c.? _____

27. If 1-inch by 8-inch bevel siding is applied to a building with 6¾ inches exposed to the weather, what height will be covered with 15 strips? _____

28. In a flight of stairs there are 13 treads. Each tread has a run of 9¼ inches. What is the total run for the flight of stairs? _____

29. A barn loft with an area of 4,254 square feet requires a matched floor. 1-inch by 6-inch T and G flooring is used, with a 1-inch overlap for each board. How much additional stock is needed to allow for overlap? _____

30. What length of each kind of stock is needed to fill the order in the chart below? (Remember that the number of saw cuts is always one less than the number of pieces needed.)

Material	Quantity	Length	Allowance for Each Saw Cut	Total Length Required
a. Bed Molding	10	8½"	¹⁄₁₆"	
b. Door Casing	4	30⅜"	⅛"	
c. Dowel Rod	32	1⅛"	¹⁄₁₆"	
d. Quarter Round	6	15¼"	⅛"	
e. Cove Molding	9	10⅞"	⅛"	
f. Screen Molding	5	3¹⁄₁₆"	³⁄₃₂"	

31. Each rail of the fence shown in 5½ inches wide. Find the total height of the fence. _____

¾"

8"

8"

10"

 Unit 9 *DIVISION OF*
 COMMON FRACTIONS

BASIC PRINCIPLES OF DIVISION

The operation of fraction division is very similar to multiplication. When dividing a common fraction, invert the divisor and proceed as in multiplication.

Example: Divide ⅓ by ½

$$\frac{1}{3} \div \frac{1}{2} = \frac{1}{3} \times \frac{2}{1}$$ Invert the divisor and change ÷ to ×

$$\frac{1}{3} \times \frac{2}{1} = \frac{2}{3}$$ Multiply numerators and denominators

When dividing fractions and whole numbers or mixed numbers, change the entire problem to fractions and proceed as above.

Examples: $$4 \div \frac{1}{3} = \frac{4}{1} \div \frac{1}{3} = \frac{4}{1} \times \frac{3}{1} = \frac{12}{1} \text{ or } 12$$

$$4\frac{1}{2} \div 1\frac{1}{4} = \frac{9}{2} \div \frac{5}{4} = \frac{9}{2} \times \frac{4}{5} = \frac{36}{10} = \frac{18}{5} = 3\frac{3}{5}$$

When multiplying any number by a number greater than 1, the product is larger than the original number. When dividing any number by a number greater than 1, the answer is smaller than the original number.

$$12 \times 3 = 36 \qquad 12 \div 3 = 4$$
(Decreased) (Increased)

When multiplying or dividing by a number less than 1, the results are reversed.

$$12 \times \tfrac{1}{2} = 6 \qquad 12 \div \tfrac{1}{2} = 24$$
(Decreased) (Increased)

PRACTICAL PROBLEMS

Divide the following quantities:

1. ¾ ÷ 3 _____

2. 3⅗ miles ÷ 3 _____

3. 4⅓ yards ÷ ⅛ _____

4. ¾ ÷ ⅞ _____

5. ⅝ ÷ ¾ _____

6. 3 inches ÷ ⅜ inch _____

7. $\frac{7}{8} \div \frac{3}{8}$ _____ 9. $5\frac{3}{8} \div 4\frac{1}{8}$ _____

8. $3\frac{3}{8} \div 6$ _____ 10. $22\frac{1}{8} \div \frac{1}{2}$ _____

Note: Before dividing denominate numbers, express both terms using the same units.

Example: $5'\,4\frac{1}{2}" \div 1\frac{1}{2}" = 64\frac{1}{2}" \div 1\frac{1}{2}" = \frac{129}{2} \div \frac{3}{2} = \frac{129}{2} \times \frac{2}{3} = \frac{258}{6} = 43"$

11. To make a flower garden fence, several pieces ¾ inch by 11½ inches are needed. How many can be cut from a piece ¾ inch by ½ inch by 16 feet? (Make no allowance for new kerfs.) _____

12. When laying 3¼-inch wide flooring in a 13-foot by 15-foot room, how many courses will be required for the 13-foot width? _____

13. A line is drawn so that ½ inch represents 1 foot. How many feet are represented by a line 4 inches long? _____

14. Determine the number of pieces of ⅜-inch plywood in a stack 30 inches high. _____

15. The distance from the surface of the first floor to the surface of the second floor measures 97½ inches. How many steps will there be if each step is 8⅛ inches? _____

16. How many pieces of floor covering 2⅜ inches wide are needed to cover 64⅛ inches of floor width? _____

17. How many boards 4⅝ inches wide does it take to cover a surface 222 inches wide? _____

18. A board 10 inches wide is ripped into strips 2⁵⁄₁₆ inches wide. If a total of ¾ inch is allowed for milling the strips, how many strips can be ripped? _____

19. How many boards, each 2⅓ feet long, can be cut from a length of stock 14 feet long? Make no allowance for saw kerfs. _____

20. How many pieces of ½-inch plywood are there in a stack 42 inches in height? _____

21. If ¼ inch represents 1 foot on a drawing, how many feet are represented by 10⅛ inches? _____

22. How many 7½-inch risers are there in a flight of stairs 105 inches high? _____

23. If studs are placed every 1⅓ feet, how many are needed for a wall 288 inches long? (Add one to start the wall.) _____

24. Siding (called 10-inch) measures 9¼ inches in width and will be laid 8½ inches to the weather. How many boards will be required for a wall 187 inches high? _____

25. The landing of a stairway is 59 inches high, and each rise is 7⅜ inches high. How many risers does it take to reach the landing? _____

26. The distance from floor to floor on a two-story house is 9 feet 8 inches. How many risers are there in the stairway if the plans call for a 7¼-inch rise for each riser? _____

27. A bookcase is 68¼ inches high. The five shelves, each 3.4 inch thick, are equally spaced. Determine the distance between the shelves if the top of the lowest shelf is 4 inches form the floor.

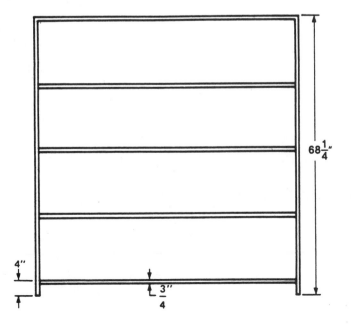

$68\frac{1}{4}''$

4''

$\frac{3''}{4}$

28. What is the height of each riser in a stair with 14 risers and a total rise of
 103¼ inches? _____

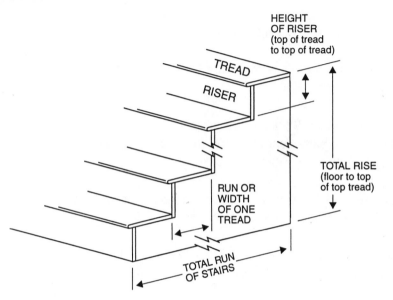

29. How many treads are needed for a set of stairs with a total run of 161¼
 inches when each tread is 10¾ inches wide? _____

Unit 10 COMBINED OPERATIONS WITH COMMON FRACTIONS

PRACTICAL PROBLEMS

Review and apply the principles of addition, subtraction, multiplication, and division of common fractions.

1. A 500-foot roll of 151-pound slater's felt is used to repair a shed roof before laying new shingles. If ¼ of the roll is used, how many feet remain? _____

2. How many 6¾-inch lengths can be cut from an 8-foot pole? (Make no allowance for saw kerfs.) _____

3. Two rooms are each 12 feet by 14½ feet, and a third room measures 10⅓ feet by 8 feet. How many square feet of hardwood flooring will be required for the three rooms? _____

4. In problem 3, assume that hardwood flooring may be ordered only in units of 50 square feet. How many units must be ordered? _____

5. Masons are paid $16.40 per hour for a 40-hour week and time and one-half for overtime. How much will a mason be paid for working 44 hours? _____

6. Three boys are hired as part-time helpers to pick up after carpenters have completed a construction job. They work 2½ hours, 1¾ hours, and 3⅓ hours every day for 5 days. What is the total number of hours all three boys work? _____

7. A clothes dryer, a washing machine, and a dishwasher are to be installed under a counter when remodeling a kitchen. Dimensions are given on the plans below. Find the overall length of the counter. _____

Decimal Fractions

SECTION

3

Unit 11 *ADDITION OF DECIMAL FRACTIONS*

BASIC PRINCIPLES OF ADDITION

Decimal fractions represent fractions with denominators that are powers of ten.

Examples: $\dfrac{3}{10}$ $\dfrac{7}{100}$ $\dfrac{13}{1000}$

When these fractions are written in decimal form, they are called decimal fractions or decimals.

$$\frac{3}{10} = 0.3 \quad \frac{7}{100} = 0.07 \quad \frac{13}{1000} = 0.013$$

To add decimals, arrange the numbers with the decimal points directly under each other and locate the decimal point for the sum under the other decimals.

Example:
$$
\begin{array}{r}
0.3 \\
4.7 \\
2.31 \\
+\ 0.08 \\
\hline
7.39
\end{array}
$$

PRACTICAL PROBLEMS

1. Add 0.325 meter and 0.05 meter. _____

2. Add 0.857 inch and 0.643 inch. _____

3. Add 4.152 centimeters, 31.648 centimeters, 6.52 centimeters, and 2.48 centimeters. _____

4. Add 0.0652 inch, 0.3148 inch, 0.5624 inch, and 0.6876 inch. _____

5. Add $33.81, $25.69, $13.72, $47.67, and $54.60. _____

6. A table top 1.857 inches thick is covered with laminated plastic 0.0625 inch
 thick. What is the total thickness of the top? _____

7. A piece of plywood has an inner core 0.325 inch thick. It is covered by two
 pieces, each 0.275 inch thick. What is the total thickness? _____

8. Find the thickness of the plywood block from the information given in the
 illustration. _____

9. Costs for framing six small buildings are $13,152.46; $14,037.92;
 $14,987.78; $13,490.12; $15,048.61; and $15,409.73. What is the total
 cost? _____

10. A contractor pays $1,250.45 for frame lumber, $650.34 for windows, and
 $156.20 for trim. What is the total cost? _____

11. A homeowner buys the following materials: lumber, $31.65; hardware,
 $4.20; paint, $16.85; and molding, $2.75. What is the total cost? _____

12. A homeowner pays a carpenter $248.00 for labor, $196.48 for shingles, and
 $6.72 for nails. What is the total cost of labor and material? _____

13. What is the total cost of the following order: crown molding, $18.20; base
 shoe, $8.76; chair rail, $26.15; and quarter round, $12.76? _____

14. A carpenter pays $120.50 for shingles, $8.25 for nails, and $112.00 for
 labor. What is the total amount paid? _____

15. A carpenter receives $1,460.25 for one job and $575.40 for another job. How much is received for both jobs? _____

16. A contractor submits a bill of $1,906.42 for framing material and trim, $65.50 for hardware, $462.00 for masonry, $170.35 for painting, and $850.67 for labor. What is the total amount for the items listed? _____

17. Records show that the materials on a job cost $1,289.45, the labor $678.92, and the overhead $69.45. The contractor wants to realize a profit of $407.56. Find the total cost to the owner. _____

18. A carpenter working on three jobs collects $962.36 for the first job, $1,443.42 for the second, and $621.92 for the third. How much is collected for the three jobs? _____

19. The bracket shown must be constructed for a specific job. Find the minimum width of material needed to make a pattern for this bracket. _____

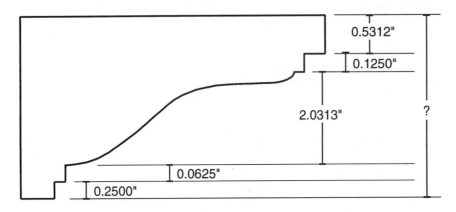

20. In bidding for a job, a contractor lists the following items: material, $1,257.45; labor, $928.75; trucking the equipment to the job, $36.20; overhead, $349.75; and profit, $270.50 What is the total bid to be submitted? _____

21. Find the total cost of the following hardware bill:

Nails	$ 16.40
Locks	29.44
Drawer pulls	5.40
Cupboard catches	3.00
Hinges, 2" × 2" butts	4.75
Bolts, carriage	4.34
Screening	21.89
Door stop	1.75
Coat hooks	3.60
Handles	1.65
Elbow catches	0.89

22. An expense record is as follows: material, $1,349.20; labor, $897.65; equipment rental, $375.00; and overhead expenses, $275.45. What are the total expenses?

23. The diagram shows a custom-made adapter shaft. A carpenter is to construct a cabinet to house this adapter shaft for reasons of safety. If the adapter shaft requires at least 1 inch of clearance on each end, what is the minimum interior length, in inches, of the cabinet?

Unit 12 SUBTRACTION OF DECIMAL FRACTIONS

BASIC PRINCIPLES OF SUBTRACTION

In order to subtract decimal fractions, arrange the numbers with the decimal points directly under each other. Locate the decimal point for the difference under the other decimals, then subtract.

Example:
```
  392.47
−  41.29
  351.18
```

When subtracting decimal fractions, it is sometimes necessary to include zeros as placeholders.

Example: 84.65 becomes 84.650
```
− 63.523              − 63.523
```

PRACTICAL PROBLEMS

1. $4,326.75
 − 2,138.47 _____

2. $2.703
 − 1.937 _____

3. $1.23 − $0.68 _____

4. 4,327.5 − 13.6 _____

5. 0.326 − 0.187 _____

6. Subtract $68.37 from $91.41. _____

7. Subtract 1.35 gallons from 2.65 gallons. _____

8. Subtract 0.365 from 0.635. _____

9. Subtract 1.87 feet from 4 feet. _____

10. Subtract 143.5 miles from 156.6 miles. _____

11. A contract is accepted for $25,050.00. The total cost of material and labor is
 $22,709.79. How much is the profit on the contract? _____

12. A carpenter accepts a job for $575.00. Bills for materials and labor are
 $113.52, $287.61, and $78.92. What is the profit? _____

13. A carpenter receives $1,400.00 for building a deck. Materials cost $671.54.
 How much is received for labor and profit? _____

14. A carpenter pays $1,647.82 for material and receives $2,567.00 for a job.
 Assuming no other costs, what is left for labor and profit? _____

15. A carpenter receives $421.25 for laying finish floor. The nails and flooring
 cost $257.60. How much is left for labor and profit? _____

16. A carpenter estimates the cost of purchasing and installing cabinets at
 $1,170.95. If the price of the cabinet is $751.45 delivered at the job
 complete, what is the cost of the labor and profit? _____

17. An estimate of $1,250.75 is submitted for a job. The actual material cost is
 $495.42, and the labor and overhead come to $596.36. What profit does the
 contractor expect on the job? _____

18. The labor for milling some stock totals 34.75 hours and is charged to three
 different jobs. To the first job, 24.5 hours are charged and to the second job,
 3.25 hours are charged. How much time is charged to the third job? _____

19. In a mill, four jobs total 146 hours. Of the 146 hours, 26.75 are charged to
 job 1; 46.25 to job 2; and 39.25 to job 3. The carpenter does not have the
 record of hours on job 4. How many hours are charged to job 4? _____

20. Find the dimension A in the decimal pattern illustrated. Express your answer in decimal terms. _____

21. With the aid of the Table III, "Metric and Customary Decimal Equivalents for Fractional Parts of an Inch" (Table III in the Appendix), convert the dimensions in the illustration to fractions and express the answer as a fraction. Finally, convert your answer to problem 20 to a fraction. Does this new answer to problem 20 agree with your answer to problem 21? _____

22. A wall mirror costs $61.70. With beveled plate glass, it costs $92.25. What is the additional cost of the beveled plate glass? _____

23. A contractor receives $231.77 for doing a piece of work. Expenses are $32.46, $35.88, and $129.68. How much profit is realized on this job? _____

24. The labor for a job amounts to $560.40. The total cost to the owner is $932.70. What is the cost of materials? _____

25. A carpenter and apprentice work a total of 76.4 hours on a job. If the carpenter's time is 38.5 hours, how much time does the apprentice work? _____

26. An estimate of $181.75 is submitted for a job. If $44.25 is for labor, what is the cost of materials? _____

27. If 14.75 squares of shingles are delivered to a job and 12.25 squares are used, how many squares are left over? _____

28. What is the length of the column in this illustration, including top and bottom plates? _____

29. An estimate for interior finish is as follows:

Doors	$162.78	Interior trim	$84.73
Windows	97.22	Shelving	27.69
Casement sash	26.38	Drawers	18.52

When the job is complete, the painter receives a credit of $14.27 on the doors, $3.25 on the sash, and $6.29 on the interior trim. Additional materials are ordered costing $6.26 and $3.75. What is the net amount of the bill? _____

Unit 13 MULTIPLICATION OF DECIMAL FRACTIONS

BASIC PRINCIPLES OF MULTIPLICATION

The numbers in a multiplication problem all have names.

$$
\begin{array}{r}
312. \quad \longleftarrow \text{Multiplicand} \\
\times .03 \quad \longleftarrow \text{Multiplier} \\
\hline
936 \\
000 \\
\hline
09.36 \quad \longleftarrow \text{Product}
\end{array}
$$

Decimal fractions are multiplied in the same manner as whole numbers. The number of decimal places to the right of the decimal point in multiplier and multiplicand are added. This total is the number of places to the right of the decimal point in the product.

Examples:

$$
\begin{array}{r}
43.21 \\
\times\ 5.42 \\
\hline
8642 \\
17\ 284 \\
216\ 05 \\
\hline
234.1982
\end{array}
\qquad
\begin{array}{r}
.004 \\
\times\ .021 \\
\hline
004 \\
008 \\
\hline
.000084
\end{array}
$$

PRACTICAL PROBLEMS

1. Multiply 1.75 by 2.95. _____

2. Multiply 2.357 by 2.57. _____

3. 0.0389 × 0.0056 _____

4. 0.947 × 0.004 feet _____

5. 1,000 × 0.365 _____

6. A builder drives 18.4 miles round trip to and from a house under construction. If 117 days are spent working on this house, what is the total mileage? _____

7. If one square foot of wall requires 6.25 firebricks, find the approximate number needed for 33 square feet. _____

8. A board is divided into 8 equal segments. If each segment is 12.250 inches long, what was the total original length of the board? (Make no allowances for saw kerfs.) _____

9. Laying fiberglass roofing on an irregular roof is estimated to take 2.6 hours per square. How long will it take to lay 14.4 squares? _____

10. The material needed for a deck is estimated to be 81.9 board feet, at a cost of $3.15 per board foot. What is the total cost of the material? _____

11. The fitting of a casement sash of a certain size takes an average of 0.375 hour. How many hours will it take to fit 26 sashes? _____

12. A carpenter can lay 100 square feet of deadening felt over a subfloor in 0.4 hour. How much time does it take to lay 5,525 square feet? _____

13. The illustration shows the floor plan of a reception room in a motel. The floor is to be covered with prefinished block flooring. If it takes 1.3 blocks to cover 1 square foot of floor, find the number of blocks needed to cover the entire floor. _____

 Note: *Area of a rectangle = length × width.*

14. Finish flooring is estimated to 3.2 hours per 100 square feet (one square). At that rate, how much time does it take to lay a floor 15.3 squares in area? _____

Unit 14 DIVISION OF DECIMAL FRACTIONS

BASIC PRINCIPLES OF DIVISION

The numbers in a division problem all have names.

Example: $20 \div 0.4 = 50$

$$
\begin{array}{r}
50 \leftarrow \text{quotient} \\
\text{divisor} \longrightarrow 0.4\overline{)20} \leftarrow \text{dividend}
\end{array}
$$

Before dividing, move the decimal point in the divisor to the extreme right, making the divisor a whole number.

$$0.4\overline{)20}$$

Move the decimal point in the dividend to the right the same number of places. Place zeros after the dividend if necessary.

$$4.\overline{)20.0}$$ 1 zero added

Place a decimal point for the quotient directly above the new decimal point in the dividend.

$$4.\overline{)200.}$$

Divide as with whole numbers.

Examples:

$$
\begin{array}{r}
50. \\
4\overline{)200.} \\
\underline{20} \\
0 \\
\underline{0}
\end{array}
\qquad
\begin{array}{r}
3.471 \\
.21.\overline{).72.891} \\
\underline{63} \\
98 \\
\underline{84} \\
149 \\
\underline{147} \\
21 \\
\underline{21}
\end{array}
\qquad
\begin{array}{r}
700. \\
.04.\overline{)28.00.} \\
\underline{28} \\
00 \\
\underline{00} \\
00 \\
\underline{00}
\end{array}
$$

PRACTICAL PROBLEMS

Note: Round answers to number of decimal places specified.

1. Divide 7.4 inches by 6. (3 places) _____

2. Divide 16 meters by 0.57. (2 places) _____

3. Divide 8.57 by 1.52. (3 places) _____

4. Divide 5.652 by 0.0652. (3 places) _____

5. Divide $95.63 by 3.5. (2 places) _____

6. Divide 64 pounds by 0.32. _____

7. Divide 64 by 1.6. _____

8. Divide 0.208 liter by 0.13. _____

9. Divide $4 by $0.08. _____

10. Divide 0.4428 by 246.0. (4 places) _____

11. Divide 6.4 by 1. _____

12. Divide 64 grams by 0.1 gram. _____

13. Divide 64 by 0.001. _____

14. Is the result in division always a number that is smaller than the quantity
 being divided? _____

15. In the figure, each of 21
 courses of siding is equally
 exposed to the weather. How
 many inches on each piece of
 siding are exposed to the
 weather?

94.50"

SIDING

16. A worker's paycheck for a job is $390. This is one and one-half times as much as the helper's check. How much does the helper receive? (Hint: Express one and one-half as a decimal fraction.)

17. A painter spends 12 hours (720 minutes) staining 1,500 square feet on a shingled building. How many minutes will it take to stain 100 square feet?

18. Vinyl to finish a kitchen floor costs $1,167.60 for labor and material. What is the cost per square foot if the area of the floor is 168 square feet?

19. How many pieces of 2.25-inch face flooring are required for a closet 34.75 inches by 6 feet when the flooring is laid parallel to the length of the closet?

20. Determine distance A in order to locate the centers for drilling two additional holes in the pattern shown.

Unit 15 EXPRESSING COMMON FRACTIONS AND MIXED NUMBERS AS DECIMALS

BASIC PRINCIPLES OF CONVERSION

To change a fraction to a decimal, divide the numerator by the denominator.

Example: Change ⅜ to a decimal.

$$
\begin{array}{r}
.375 \\
8\overline{)3.000} \\
2\,4 \\ \hline
60 \\
56 \\ \hline
40 \\
40 \\ \hline
\end{array}
$$

Those fractions yielding a repeating decimal should be rounded off to the desired number of places.

Example:
⅓

$$
\begin{array}{r}
.3333 \\
3\overline{)1.0000}
\end{array}
$$

⅓ = .33333333 . . .
Round off to 2 places = .33

In order to express decimals as fractions, drop the decimal point and write the given number as the numerator. Write the denominator as a power of ten (10, 100, 1,000, etc.), using as many zeros as there are decimal places in the decimal number.

Examples: $.31 = \dfrac{31}{100}$ $.025 = \dfrac{25}{1000}$ $5.4 = 5\dfrac{4}{10}$ or $\dfrac{54}{10}$

PRACTICAL PROBLEMS

Express the common fractions as decimal fractions in problems 1 through 4, and decimal fractions as common fractions in problems 5 through 8. Refer to the table of decimal equivalents for fractional parts of an inch in the Appendix (Table III).

1. ⅜ _____

2. ¾ _____

3. ¼ _____

4. ²⁷/₃₂ (3 places) _____

5. 0.78125 _____ 7. 0.53125 _____

6. 0.6875 _____ 8. 0.546875 _____

9. Approximately 1¼ pounds of nails are needed for each 100 square feet of subflooring. Write this weight in decimal form. _____

10. The actual width of a pine board is 7¼ inches. Write the width in decimal form. _____

11. The actual thickness of a piece of textured panel is ⁷⁄₁₆ inch. What is the thickness written as a decimal fraction? _____

12. A piece of plywood is 0.625 inch thick. What is its thickness in common fraction form? _____

13. Find the approximate thickness in common fraction form of a piece of siding 0.4375 inch thick. _____

14. A floor has a concrete slab with an elevation of 113.45 feet. It is to have two cement finishes with a total thickness of ¾ inch. What is the elevation of the finished floor?

 a. Express ¾ inch as a decimal fraction. a. _____
 b. Express the answer to part (a) as a fraction of a foot in decimal form. b. _____
 c. Find the total elevation of the finished floor in feet. c. _____

Note: Use this illustration for problems 15 and 16.

2" TERRAZZO FINISH
⅝" PARTICLE BOARD
½" PLYWOOD
HANGER
FLUSH TYPE GIRDER
FLOOR JOIST

15. If the elevation to the top of the girder shown is 171.46 feet, find the elevation in feet to the top of the particle board subfloor. Proceed as follows:

 a. Express ½ inch as a decimal. a. _____
 b. Express the result as a decimal fraction of a foot by dividing it by 12. Round to 3 decimal places. b. _____

 c. Express ⅝ inch as a decimal.

c. _____

 d. Express result as a decimal fraction of a foot by dividing it by 12. Round to 3 decimal places.

d. _____

 e. Add all decimal fractions of a foot.

e. _____

16. If the top of the girder is 156.92 feet in elevation, find the elevation in feet to the top of the particle board subfloor. Follow the steps used in problem 15.

a. _____

b. _____

c. _____

d. _____

e. _____

Note: To solve problems 17–20, use a method similar to problem 16.

17. In setting form work for a sidewalk, a carpenter is asked to raise the form ¾ inch above an elevation of 141.35 feet. What is the new elevation?

Note: Use this illustration for problems 18–20.

18. What is the elevation at the top of the concrete setting bed if the elevation is 146.305 feet at the top of the tile floor?

19. What is the elevation at the top of the ⅞-inch plywood floor in the figure?

20. What is the elevation at the top of the girder in the figure?

21. Express 0.625 inch as eighths of an inch.

22. The illustrations below represent segments of a carpenter's rule. Working directly from the rule, at each arrow express the reading as its decimal equivalent.

A _____

B _____

C _____

D _____

E _____

F _____

G _____

Note: Builders often use an instrument called a *builder's level* to measure vertical distances. The instrument is focused on a graduated stick called a Philadelphia rod with markings divided into feet and decimal parts of a foot. The first sighting is on a measurement above a fixed reference point called a *benchmark*. The measurement above the benchmark is called the *height of instrument*.

23. What is the height of instrument? _____

24. What is the elevation at point A? _____

25. What is the elevation at point B? _____

26. What is the difference in elevation between points A and B? _____

Note: The Philadelphia rod has large numbers to indicate whole feet, smaller numbers to indicate tenths of a foot, and unnumbered graduations to indicate hundredths of a foot. The markings and spaces are equal in size, so that readings are taken from the top or bottom of the marking. For problems 27 through 33, write the rod readings in whole feet and decimal parts (hundredths) of a foot.

Note: Use this illustration for problems 27–33.

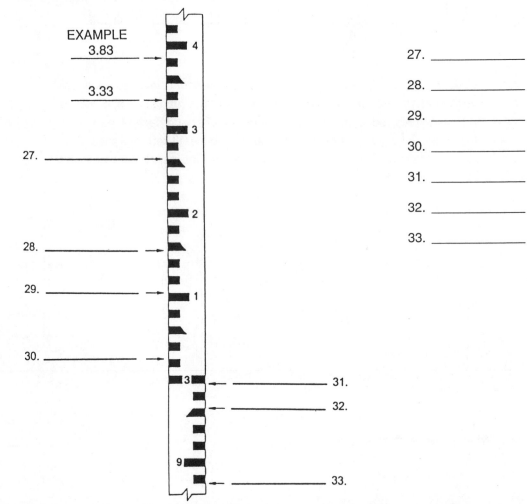

27. _____

28. _____

29. _____

30. _____

31. _____

32. _____

33. _____

Percent and Percentage

Unit 16 SIMPLE PERCENT AND PERCENTAGE

BASIC PRINCIPLES OF PERCENT

Percent means *each hundred*, just as miles per hour means miles each hour. ("Cent" refers to hundred, as in *cent*ury or *cent*ipede). Forty miles per (one) hour can be expressed as the fraction $^{40}/_1$; 25% can be expressed as $^{25}/_{100}$. Percent may be expressed as a decimal by first changing the number to a fraction and then to a decimal.

Example: $37\% = \dfrac{37}{100} = 0.37$

The % sign may be considered a magnet that attracts the decimal point two places to the right. Putting % on a decimal number pulls the decimal point two places to the right. Removing the % sign moves the decimal point two places to the left.

Examples: 0.81 = 81.% or 81% 37.% = 0.37

When solving percentage problems, first change the percent to a decimal, and then treat "of" as "times."

Examples: Find 3% of $75.
3% of $75 = 0.03 × $75 = $2.25

Find 150% of 400 lb.
150% of 400 lb = 1.50 × 400 lb = 600 lb

A basic formula for percentage problems is:

$$Part = Percent \times Whole$$
$$P = \% \times W$$

The following diagram may help in using this formula.

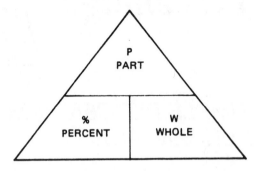

To use this diagram, cover the section you are solving for to determine the formula to use. For example, when determining Percent (%), cover the percent section of the diagram; the fractional parts P (Part) and W (Whole) remain. Therefore, the formula to use to find Percent is:

$$\% = \frac{P}{W}$$

Example: What percent of 28 is 7?

$$\% = \frac{P}{W}$$

$$\% = \frac{7}{28} = 0.25 = 25.\%$$

$$\% = 25$$

When determining the Whole, cover W in the diagram; P and % remain. The formula to find the Whole is:

$$W = \frac{P}{\%}$$

Example: 3 is 25% of what number?

$$W = \frac{P}{\%}$$

$$W = \frac{3}{0.25} = 12$$

$$W = 12$$

PRACTICAL PROBLEMS

1. The Whole is also the Base (*B*); the Percent is also the Rate (*R*); and the Part is also the Percentage (*P*). Use the formula **Percentage = Base × Rate** to find the following:

 a. $P = 493$; $R = 85\%$; $B = ?$ a. _____

 b. $P = ?$; $R = 55\%$; $B = 125$ b. _____

 c. $P = \$588$; $R = ?$; $B = \$840$ c. _____

 d. $P = ?$; $R = 22\%$; $B = 416.8$ d. _____

 e. $P = 240$; $R = ?$; $B = 192$ e. _____

2. In the figure shown, what percent of board width is allowed for matching? (The difference between the two dimensions shown is allowed for matching.)

Note: To determine the amounts of material needed for various jobs, it is a common practice to allow a percentage for waste in cutting.

3. It is estimated that 2,550 square feet of roof sheathing is needed. If an additional 20% is allowed for waste, find the total amount required. _____

4. How many square feet of floor felt should be ordered for a job when the area of the floor is 1,540 square feet and 15% of the area is allowed for waste? _____

5. The total area to be covered with insulation board is 1,250 square feet. How much should be ordered if 8% of the area is added for waste? _____

6. It is estimated that about 65 feet of flashing material for drip caps are needed for a certain job. How much should be provided if 20% of the estimate is added to allow for waste? _____

7. In making mortar, lime is used in an amount equal to 12% of the cement. How much lime is necessary if 995 pounds of cement are used? _____

8. A floor of 385 square feet is to be covered with subflooring laid diagonally. To allow for waste, 25% of the area is added when subflooring is laid diagonally. How many square feet of subflooring should be ordered? _____

9. It is estimated that 10,050 face bricks are required for a certain job. How many more bricks are required if 14% of the estimate is added for waste? _____

10. A carpenter estimates a need for 280 board feet of flooring but forgets to allow for waste. He then adds 25% to this amount. What is the total number of board feet in the carpenter's order? _____

11. Not allowing for waste, a carpenter estimates that 728 board feet of matched flooring are needed for a particular job. If 20% is added for waste, what is the waste allowance for this job? _____

12. A carpenter must order subflooring for 2 floors of the same size. Without allowing for waste, the carpenter estimates that each floor requires 1,080 feet. If 25% is allowed for waste, how many board feet of subflooring are ordered? _____

13. In constructing a building, 2% of the total cost is allowed for excavation. Find the cost of the house if the excavation costs $825. _____

14. The carpentry and millwork on a new structure amounts to $36,756. A contractor figures 48% of this amount as the cost of material and 32% as labor.

 a. What is the material cost? a. _____

 b. What is the labor cost? b. _____

15. A contractor receives $1,800 profit on a job. If this is 12% of the total charge, what does the owner pay? _____

16. A general contractor estimates a nonresidential building to cost $125,860. If the work of excavating and grading is 3% of this amount, the concrete work is 20%, and the carpentry is 9%, what is the amount estimated for each of these three items?

excavating and grading = _____

concrete work = _____

carpentry = _____

17. The total cost of building a house is $86,000.00. What percent is spent on carpentry labor and material if the total for these costs is $42,275.50? (Express answer to nearest tenth percent.)

18. On a nonresidential building, 6% of the total job cost is carpentry labor. If the total cost is $350,000, what is the cost for carpentry labor?

19. The carpentry labor and material on a building is estimated to be $48,500. Of this amount, 37% is for labor and 45% is for materials.

a. What is the allowance for labor?

a. _____

b. What is the allowance for materials?

b. _____

c. How much is left for overhead and profit?

c. _____

20. The grading on a $45,500.00 house comes to $787.50. What percent of the total cost is this? (Express answer to nearest hundredth percent.)

21. A contractor purchases 17,500 board feet of rough lumber. If there is waste of 18% in milling, how many board feet of lumber are actually received?

22. A certain grade of lumber is supposed to be 75% clear, or free from knots and other defects. How much clear lumber should a carpenter expect to find in a load of 2,200 board feet?

23. A dealer buys an item at a cost of $6,600 and sells it at a profit equal to 12% of the cost. Find the amount of profit and the selling price.

Note: *Selling price = Cost + Profit.*

24. A contractor pays workers $68.80 per day for labor. If the contractor makes a 10% profit, find the contractor's profit and the charge to the customer for the workers' labor.

Note: *Charge to customer = Contractor's cost + Contractor's profit.*

25. How much should a contractor bid for a job if costs total $2,300 and a profit of 12% is desired?

 Note: *Amount of bid = Cost + Profit.*

26. A carpenter charges a customer $862.50 for a job. This includes a profit of $112.50. What is the carpenter's rate of profit?

27. What is the bid on a job with estimated costs of $3,464 if the profit made is 11% of the cost?

28. The estimated cost of a small house totals $79,375.00. The contractor adds 8% for profit. What is the exact amount of the bid?

29. A contracting company submits a bid for an apartment house. The estimated cost is $181,693.21. To this, add 23% of the estimated cost for overhead, and 9% of the estimated cost plus the overhead for profit. What is the final bid?

30. To an estimate of $43,568.84 a contractor adds 2% for incidentals, and 8% of the estimate plus incidentals for profit. What is the total bid?

31. A general contractor's bid shows the following net costs and profits (percent of costs) added by subcontractors:

Plumbing	$4,528.10	+	7%
Masonry	$6,278.43	+	11%
Electrical work	$1,829.71	+	9%
Heating	$2,372.17	+	6%

 In addition to these costs, the general contractor estimates costs of $61,264.38, and adds a further 12% of the total for profit. What is the general contractor's total bid?

32. What percent of the lot illustrated is taken up by the house and driveway? _____

 Note: *Area of rectangle = length × width.*

- 150'

HOUSE =
2,150 SQUARE
FEET

90'

DRIVE = 820 SQUARE FEET

Unit 17 INTEREST

BASIC PRINCIPLES OF INTEREST

When a person borrows money, a fee for the use of that money is usually charged by the person lending the money. An example of this is the finance charge that is added for buying items on credit. An amount, in addition to the cost of the item, is charged the customer; this is a form of interest. Banks also pay interest to their depositors for the privilege of using the depositors' money. A carpenter or developer often borrows money to build a house. Interest must be paid on the amount borrowed until the house is sold and the loan can be repaid. Interest is usually expressed as a percent of the principal.

Study the following definitions that relate to computing interest:

Principal is the amount of money invested or borrowed.
Interest is a payment for the use of money; it is expressed as a percent of the principal.
Rate of interest is the percent at which the interest is calculated.
Amount includes principal plus interest.
Compound interest is interest based on the principal plus previously earned interest.

FORMULAS

1. *Simple interest = Principal × Rate × Time* in years. In symbols, this is $I = PRT$.

2. *Amount = Principal + Interest*

Examples:

1. Find the simple interest and amount on $800 at 12% after 2 years

 Interest = *Principal × Rate × Time* (in years)
 Interest = $800 × 12% × 2
 = $800 × 0.12 × 2
 = $192

 Amount = *Principal + Interest*
 Amount = $800 + $192
 = $992

2. The annual rate of interest on $1,200 is 14%. Find the interest after a period of 3 months.

 Note: 3 months = $\frac{3}{12}$ year = $\frac{1}{4}$ year.

 $Interest$ = $Principal \times Rate \times Time$
 $Interest$ = $\$1,200 \times 0.14 \times \frac{1}{4}$
 = $\$42$

PRACTICAL PROBLEMS

Determine the simple interest and amount under each of the following conditions.

	Principal	Rate of Interest	Time	Interest	Amount
1.	$ 300	9%	4 years	_____	_____
2.	$ 640	11%	½ year	_____	_____
3.	$ 4,280	12%	3 months	_____	_____
4.	$ 20,000	15½%	1 year	_____	_____
5.	$ 1,080	9½%	1½ years	_____	_____

6. A family pays 9% interest on a $95,000 mortgage for the construction of a house. What is the yearly interest paid on the mortgage? _____

7. A carpenter deposits $192.80 in a savings bank. If the bank pays 2¼% interest per year, payable annually, what is the total amount in the account at the end of one year? _____

8. A bank loans $44,250 to a contractor at a rate of 9% per year. What is the yearly interest payment? _____

9. A contractor pays $450 interest annually on a mortgage of $3,600. What is the rate of interest? _____

10. A contractor borrows $30,000 on a building loan at 9½% annual interest. What does the loan money cost for 1 year? _____

11. A builder develops a tract of 20 houses at a cost of $86,000 each. If the mortgage is placed on one-half of the total cost of the 20 houses at 8¾% interest, what interest is paid each year? _____

12. Two business partners have a bank balance of $7,942.00. After the first year, their interest brings their balance to $9,053.85. What rate of interest is earned? _____

13. A contractor borrows the sum of $9,973 and pays the bank 10½% interest yearly. How much interest is paid for 1 year? _____

14. Find the interest for 1 year at 9½% on a lumber list of $3,896.75. _____

Note: In certain cases, interest is figured monthly, quarterly, or semiannually; the contract must clearly state which. The simple interest formula, $I = PRT$, is used, but T represents the *number of interest periods*, instead of the number of years. (*Interest = Principal × Rate × Time* in number of interest periods.) If interest is figured monthly, the interest period is 1 month.

Example: A bond of $32,000 carries an interest rate of 3% for one month. Find the interest for 3 months.

Interest = Principal × Rate × Time (number of interest periods)
Interest = $32,000 × 0.03 × 3 = $2,880

15. A contractor secures a bond of $125,000 at a rate of 2% for one month. If this bond is held for four months, how much does the bond cost for that period? _____

16. A contractor gives a bank a note for $7,550 at a rate of 1% for one month. How much interest is charged for 3 months? _____

17. A hardware company adds 2% to bills that are not paid within a 30-day period. A carpenter is unable to pay a bill of $70.80 until after 30 days. What is the interest charged on the bill? _____

18. A lumber yard levies a 12% interest charge (1% per month) on money owed them after 30 days. How much interest does a contractor pay on a $1,876 bill that is paid in 90 days? _____

Note: *Compound interest* is computed on the original principal *plus previously earned interest*. Interest compounded *quarterly* is added to the account every 3 months (¼ year); there are 4 interest periods each year. The interest rate for a period equals the annual *interest rate* divided by the number of interest periods in 1 year.

19. A builder makes a $1,200 profit on the sale of a new home. The profit is deposited in a bank at 3% interest, compounded quarterly. What is the bank balance at the end of 8 months if no deposits and no withdrawals are made during that time? To compute compound interest, follow these steps:

 a. How many interest periods are there in 1 year? a. _____

 b. For how many interest periods is the money left in the bank? b. _____

 c. What is the interest rate for 1 period? c. _____

 d. Find the interest for the first period. (*Principal × Rate for 1 period = Interest for 1st period*) d. _____

 e. Find the amount at the end of the first period. (*Principal + Interest = Amount*) e. _____

 f. Find the interest for the second period. (*Amount at end of 1st period × Rate for 1 period = Interest for 2nd period*) f. _____

 g. Find the amount at the end of the second period. (*Amount at end of 1st period + Interest for 2nd period = Amount at end of 2nd period*) g. _____

20. A contractor has $3,755 in a 4% interest-earning account. If interest is paid every 6 months and no withdrawals are made, how much money is in the bank at the end of 3 years? (Hint: Use the steps outlined in problem 19.) _____

Unit 18 DISCOUNTS

BASIC PRINCIPLES OF DISCOUNT

Discount rate is the percent by which a price is marked down. *Discount* is the amount by which the price has been marked down. The *list price* (L) multiplied by the *discount rate* (R) equals the *discount* (D):

$$L \times R = D$$

The list price minus the discount equals the sale price, or *net price* (N).

Example: A hammer marked $22.50 is being sold at a 10% discount. Find the sale price.

$L \times R = D$
$\$22.50 \times 0.10 = \2.25

$L - D = N$
$\$22.50 - \$2.25 = \$20.25$

Sometimes a double discount is given, which is a discount on the sale price.

Example: A panel saw lists for $400. It is on sale, marked down 10%. An additional 2% discount is given for paying cash. If a buyer pays cash, what is the cost?

$\$400 \times 0.10 = \40
$\$400 - \$40 = \$360$
$\$360 \times 0.02 = \7.20
$\$360 - 7.20 = \352.80

There is another, much faster, method of solving double discount. Think of "discount" as "don't pay." Therefore, when there is a 10% discount, you *do pay* 90%. Likewise, a 2% discount means you *do pay* 98%. This example may be solved quickly as follows:

$\$400 \times 90\% \times 98\% = \$400 \times .9 \times .98 = \$400 \times .882 = \$352.80$

PRACTICAL PROBLEMS

1. List price is $5,670; discount is 12% of this price. Find the cost. _____

2. List price is $5,780; discount is 12% of the list price; cash discount is 2%. Find the net cost. _____

3. Retail price is $58; discount is 20% of retail price; cash discount is 3%. Find the net cost. _____

4. List price is $73; discount is 10%; cash discount is 2%. Find the net cost. _____

5. An item is on sale at a 25% discount. The sale price is what percent of the list price? _____

6. Net cost is $28.80 after a discount of 20% is allowed off retail price. Find the retail price. (Hint: The net cost is 100% – 20%, or 80% of the retail price. To find the retail price, divide the net price by the 80%.) _____

7. A discount of 2% is allowed for cash. If the amount of cash paid is $1,225, find the original amount of the bill. _____

8. Cost is $15.86 after taking 35% off catalog price. Find the catalog price. _____

9. Net cost is $5,188.80 after taking 8% off list price. Find the list price. _____

10. If 2% is allowed for cash, and the amount of cash paid is $1,025, find the original amount of the bill. _____

11. Net cost is $5,831; discount on retail price is 15%; cash discount is 2%. Find the selling price. _____

12. Net cost is $630.63; discount on retail price is 10%; cash discount is 2%. Find the retail price. _____

13. If cost is $15.86 after taking 20% off the catalog price, find the catalog price. _____

14. When the amount paid is $154 after a 15% discount is allowed, determine the original amount of the bill. _____

15. List price is $48.92; net price is $44.03. Find (a) the amount of discount, and
(b) the rate of discount. (Express answer to nearest whole percent.) a. _____

 b. _____

16. List price is $792.47; net price after cash discount is $776.62. Find (a) the
discount, and (b) the rate of discount. (Express answer to nearest whole
percent.) a. _____

 b. _____

17. If a bill is paid within 10 days, the customer is charged a net total of $27.60.
If the bill is paid after 10 days, the gross of $28.45 is charged. What is the
rate of discount for prompt payment? (Express the answer to the nearest
whole percent.) _____

18. A contractor orders material that costs $1,926.90, less 2% if paid within 30
days. How much is due if the bill is paid 10 days after ordering? _____

Note: As previously noted, in addition to the single discount allowed on some
commodities or building materials, business practice often permits the
giving of a second discount. After the first discount is taken off, the second
discount is figured on the remainder. In trade parlance, the double
discount is known as 10 and 2 off, 12 and 4 off, or whatever is relevant.
In all the following problems, round your answers to the nearest whole
cent.

19. A contractor receives a large bill for hardware and is given discounts of 15%
and 5% off. What is the net cost for the hardware, the price quotation of
which is $781.60? _____

20. A hardware bill totals $345.69 with discounts of 5% and 3%. What is the net
cost of the material? _____

21. A catalog quotation on a bill for millwork amounts to $1,241.50. The firm
allows a discount of 2% and 1%. How much does a carpenter have to pay
for this material? _____

22. A tool chest containing 12 items is listed for $95. Discounts of 15% and 10%
are allowed. If the customer pays cash within 30 days, an additional
discount of 5% is granted. What is the cost of the set if a carpenter takes
advantage of all the discounts offered? _____

23. The dealer's price to a contractor on a bill for materials is 12½% off list price, with an additional 2% discount for cash within 30 days.

 a. Find net price on a $1,975 order.

 b. Find the total discount.

 a. _____

 b. _____

24. A hardware firm sells nails at a discount of 28% off the list price. What is the cost of 540 pounds of 8d nails that are listed at 51 cents per pound?

25. A contractor purchases oak lumber for $1,400.00, less 2% discount; yellow pine for $596.00, less 2%; white pine for $896.50, less 1%. What is the total amount of the bill?

Measurement: Direct and Computed

SECTION 5

Unit 19 LINEAR MEASURE

BASIC PRINCIPLES OF LINEAR MEASURE

Linear measure refers to a measure of length. Usually, this type of measurement is used to measure and order standard types of lumber. For example, studs, molding, rafters, and siding are ordered by length in linear, lineal, or running feet.

In the English system, the units of linear measure are inches, feet, yards, and miles. In the metric system, the basic unit of linear measure is the meter.

COMMON ENGLISH LINEAR UNITS		
12 inches	=	1 foot
3 feet	=	1 yard
16½ feet	=	1 rod
1,760 yards	=	1 mile
5,280 feet	=	1 mile

COMMON METRIC LINEAR UNITS	SYMBOL	VALUE IN METERS
1 millimeter	mm	0.001
1 centimeter	cm	0.01
1 decimeter	dm	0.1
1 meter	m	1.0
1 dekameter	dam	10.0
1 hectometer	hm	100.0
1 kilometer	km	1000.0

(COMPARATIVE SIZES ARE SHOWN)

1 METER

1 YARD

Meter: A little longer than a yard (about 1.1 yards or 39.37 inches)
Millimeter: 0.001 meter; diameter of a paper clip wire
Centimeter: 0.01 meter; width of a paper clip (about 0.4 inch)
Kilometer: 1,000 meters; somewhat further than ½ mile (about 0.6 mile)

Perimeter is the linear measure of the boundaries of a figure or structure.

The perimeter of a *square* is the total length of its four sides. Expressed as a formula,

$$P = 4s \text{ with } P = \text{perimeter and } s = \text{length of side}$$

To find the perimeter of a *rectangle*, use the formula

$$P = 2(w + l) \text{ with } P = \text{perimeter, } w = \text{width of rectangle, and } l = \text{length of rectangle}$$

The use of parentheses in a formula indicates that the mathematical operation within the parentheses must be performed before any other operation.

Circumference is the linear measure of the boundary of a circle.

To find the perimeter or circumference of a circular object, use the formula $C = \pi d$ or $C = 2\pi r$ with C = circumference, d = diameter, r = radius of circle, and π (pi) is a Greek letter or symbol used to compute relationships in circular measure. Its value is constant and may be expressed approximately as 3½, ²²⁄₇, 3.14, or 3.1416, depending on the degree of accuracy required.

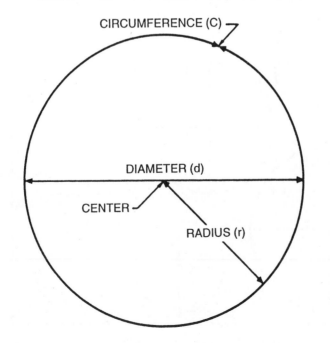

PRACTICAL PROBLEMS

1. Using a rule, measure the length of each line segment to the nearest ¹⁄₁₆ of an inch. Reduce answer to lowest terms.

a ———————————————————— a _____

b —————————————— b _____

c —————————————————————————— c _____

d ———————————————————————— d _____

e —————————————————— e _____

f —————————— f _____

g —————————————————————— g _____

h ———————————————————— h _____

i —————————————————————— i _____

2. Using either a caliper or scale, measure the diameter of each bored hole to the nearest ¹⁄₁₆ inch.

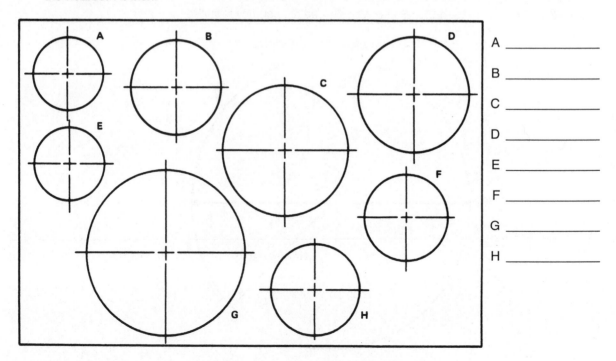

A _____

B _____

C _____

D _____

E _____

F _____

G _____

H _____

3. Using a metric scale, measure the length of each line segment, correct to the nearest millimeter.

a _____

b _____

c _____

d _____

e _____

4. Measure the diameter of each hole to the nearest millimeter.

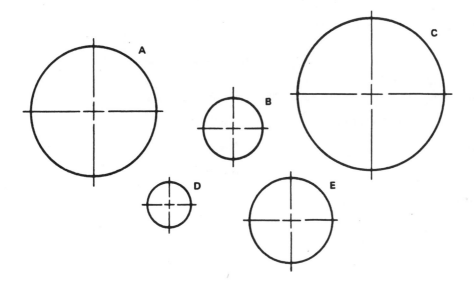

A _____

B _____

C _____

D _____

E _____

Example of finding the perimeter of a square:

Find the perimeter of a square with a side of 9 feet.

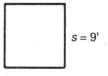

$s = 9'$

1. The formula for the perimeter of a square is $P = 4s$.

2. Substitute 9' for the symbol *s*. $P = 4 \times 9'$

3. Multiply. $P = 36'$

Find the perimeter of each of the following squares.

5. Side = 18" _____

6. Side = 34" _____

7. Side = 25 m _____

8. Side = 32' _____

9. Side = 42 km _____

10. $s = 53'$ _____

11. $s = 2' 5"$ _____

12. $s = 4' 7"$ _____

13. $s = 14' 6"$ _____

14. $s = 22' 9"$ _____

Example of finding the perimeter of a rectangle:

Find the perimeter of the illustrated rectangle.

$w = 5'$

$l = 9'$

1. The formula for the perimeter of a rectangle is $P = 2 \times (w + l)$.

2. Substitute 5' for w and 9' for l.

$P = 2 \times (5' + 9')$

3. Add, then multiply.

$P = 2 \times (14')$

$P = 28'$

Find the perimeter of each of the following rectangles:

15. Width = 11', Length = 14' _____

16. Width = 12', Length = 16' _____

17. Width = 150", Length = 220" _____

18. Width = 18", Length = 32" _____

19. Width = 21 mm, Length = 22 mm _____

20. $w = 3', l = 5'$ _____

21. $w = 5', l = 16'$ _____

22. $w = 8$ mm, $l = 14.5$ m _____

23. $w = 8$ m, $l = 14.52$ m _____

24. $w = 18.75"$, $l = 38.5"$ _____

Example of finding the circumference of a circle:

Find the circumference of a circle if the radius
is 7 feet. Use $\pi = 3.1416$.

1. The formula for the circumference of a circle is $C = 2\pi r$ or $C = \pi d$.

2. Substitute 3.1416 for the symbol π and 7' for r. $C = 2 \times 3.1416 \times 7'$

3. Multiply. $C = 43.9824'$ or $43'\ 11^{25}\!/_{32}"$

Note: The conversion of 43.9824 feet to 43 feet $11^{25}\!/_{32}$ inches is performed as follows:

43.9824 ft \times 12 in/ft = 11.7888 in
.7888 \times 32 = 25 to nearest whole number

Using $\pi = 3.1416$, find the circumferences of the following circles:

Note: Solve problems 25–29 for circumference correct to the nearest $^1\!/_{32}$ of an
inch.

25. Diameter = 32" _____ 28. Radius = 6' _____

26. Diameter = 4" _____ 29. Diameter = 22' _____

27. Diameter = 13' _____

Note: Solve problems 30–34 for circumference correct to the nearest
thousandth of the unit given.

30. $d = 16$ in _____ 33. $r = 14.5$ ft _____

31. $d = 17$ ft _____ 34. $d = 5$ yd _____

32. $d = 475$ mm _____

Note: Use this illustration for problems 35–39.

35. What is the perimeter of the living room in the plan shown? (Do not deduct for openings.)

36. What is the perimeter of the outside walls?

37. How many feet of baseboard are needed for the living room? (Deduct 3 feet for single door openings, and 5 feet for double door openings.)

38. How many linear feet of shoe base are needed for the two bedrooms? (Deduct 3 feet for each door opening.)

39. How many linear feet each of base and tile cap are used in the bathroom? (Deduct 3 feet for each door opening.)

40. What is the perimeter of the semicircular form in the figure? (Express the answer to the nearest thousandth meter.)

r = 2 m

41. A garage footing 18 inches wide measures 23 feet by 25 feet along its outside dimensions. Find the inside dimensions.

42. How many linear feet each of baseboard, base cap, and base shoe are required for a room 14 feet by 18 feet with two doors, allowing 3 feet per door?

43. A new subdivision has outside dimensions of 978 feet by 3,045 feet. What is the distance around it?

 Unit 20 WORKING WITH FEET
AND INCHES

BASIC PRINCIPLES OF FEET AND INCHES

Dimensions expressed in combined feet and inches can usually be added by simply adding the feet, adding the inches separately, and then expressing the inches as a combined value of feet and inches.

Example: Add 1 foot 7 inches and 4 feet 8 inches.

$$
\begin{array}{r}
1 \text{ foot} \quad 7 \text{ inches} \\
+ \ 4 \text{ feet} \quad 8 \text{ inches} \\
\hline
5 \text{ feet } 15 \text{ inches (1 foot 3 inches)}
\end{array}
$$

The answer is 6 feet 3 inches.

For more complex problems, it is often easier to express all dimensions in inches. To express feet and inches in inches, multiply the number of feet in the given dimension by 12 (the number of inches in a foot). To this product add the number of inches in the given dimension.

Example: Express 7 feet 9¾ inches (7' + 9¾") as inches.

7 × 12 = 84
7 feet = 84 inches
84 inches plus 9¾ inches = 93¾ inches

After all quantities to be worked with have been expressed in inches, perform the necessary operations. The result will be in inches, square inches, or cubic inches, depending on the operation. When adding or subtracting linear dimensions, the result will be in inches, which can be converted to feet and inches.

To express inches as feet and inches, divide the whole number in the given number of inches by 12 to find the number of whole feet. The remainder is the number of inches in addition to the number of whole feet. The answer is the number of whole feet plus the remainder in inches.

Example: Express 176⁷⁄₁₆ inches as feet and inches.

$$
\begin{array}{r}
14 \ \text{(feet)} \\
12 \overline{) \ 176\tfrac{7}{16}} \\
\underline{12} \\
56 \\
\underline{48} \\
8\tfrac{7}{16} \text{ remainder (inches)}
\end{array}
$$

The answer is 14 feet 8⁷⁄₁₆ inches.

PRACTICAL PROBLEMS

Express the following as inches:

1. 3 ft 8 in _____

2. 16 ft 11 in _____

3. 13' 9" _____

4. 5' 6½" _____

5. 19' 9¾" _____

6. 33' 4⅝" _____

Express the following as feet and inches:

7. 39 in _____

8. 135 in _____

9. 297" _____

10. 631½" _____

11. 174⅝" _____

Note: Use this illustration for problems 12–14.

12. How far in feet and inches is the centerline (℄) of the window from the left end of the building? _____

13. How far in feet and inches is the centerline (℄) of the window from the right end of the building? _____

14. How wide in feet and inches is Bedroom 1? _____

15. If 3½ inches are cut off one end of an 8-foot length of lumber, what length remains? _____

16. If a girder is placed in the center of a 26-foot 8-inch span, what is the dimension from one end of the span to the centerline of the girder? _____

Note: Use this illustration for problems 17 and 18.

17. What is the combined height of the wall and the sill? _____

18. What is the height of the footing? _____

19. How much would have to be cut from the end of a 12-foot piece of lumber to leave a piece 8 feet 5 inches long? _____

20. What is the perimeter of a room 12 feet 6 inches wide and 15 feet 3 inches long? _____

21. A girder 38 feet 9 inches long is to be supported by 4 steel posts spaced equally. How far apart should the posts be placed? _____

Find the perimeter of each of the following objects:

22.

23.

24.

Unit 21 SQUARE MEASURE

BASIC PRINCIPLES OF SQUARE MEASURE

Area is always measured in units of *square* feet, *square* inches, *square* yards, and so forth. One inch times one inch equals one *square* inch, one foot times one foot equals one *square* foot, and so on.

1 FT 1 FT × 1FT = 1 SQUARE FOOT

To find the area of a rectangle, the width is multiplied by the length. (All dimensions must be in the same units. Inches must not be multiplied by feet, or yards by inches, etc.)

Example: Find the area of a rectangle 18 inches wide and 3 feet long.

First change 18 inches to 1½ feet, then multiply 1½ feet by 3 feet. The answer will be 4½ *square* feet.

Common area formulas:

Square: *Area = side × side = s × s = s^2*

Rectangle: *Area = length × width = l × w*

Triangle: *Area = ½ × base × height = ½ b × h*

Circle: *Area = π × radius × radius* or *πr^2*

In the English system, area is measured in square inches, square feet, square yards, and square miles. In the metric system, the basic unit of square measure is the square meter.

ENGLISH UNITS OF SQUARE MEASURE
144 square inches = 1 square foot
9 square feet = 1 square yard
4,840 square yards = 1 acre

SQUARE MEASURE UNIT	SYMBOL	RELATION TO SQUARE METER
1 square meter	m^2	standard unit of area
1 square decimeter	dm^2	0.01 square meter
1 square centimeter	cm^2	0.0001 square meter
1 square millimeter	mm^2	0.000001 square meter

METRIC UNITS OF SQUARE MEASURE
$100 \ dm^2 = 1 \ m^2$
$100 \ cm^2 = 1 \ dm^2$
$100 \ mm^2 = 1 \ cm^2$
$1,000,000 \ mm^2 = 1 \ m^2$

PRACTICAL PROBLEMS

Note: In problems 1–8, express answers in square feet (sq ft) or centimeters squared (cm^2), as is appropriate.

Find the area of each of the following squares:

1. Side = 11' 0" _____ 3. Side = 5' 3" _____

2. Side = 29 cm _____ 4. Side = 6' 9" _____

Find the area of each of the following rectangles:

5. Width = 5', Length = 7' _____

6. Width = 9 m, Length = 17 m _____

7. Width = 4' 10", Length = 8' 2" _____

8. Width = 15' 0", Length = 36' 6" _____

Find the total wall area for each of the following rectangular rooms: (w = room width, l = room length, h = wall height. No allowance need be made for wall openings such as door, windows, etc.).

9. $w = 8'$ $l = 9'$ $h = 7'$ _____

10. $w = 6 \ m$ $l = 8 \ m$ $h = 2.5 \ m$ _____

11. $w = 18'\ 6"$ $l = 25'\ 6"$ $h = 9'\ 3"$ _____

12. $w = 6$ m $l = 10.5$ m $h = 2.9$ m _____

13. An apartment building contains 14 units, each 18 feet 6 inches by 26 feet 6 inches. What is the total floor area? _____

14. How many square feet of floor area are there in three two-story apartment houses, each of which is 38 feet wide and 76 feet long? _____

Note: Use this diagram for problem 15.

15. Both sides of the roof shown are covered with asphalt shingles. Find the number of square feet of surface covered. _____

16. A bathroom 6 feet 6 inches wide and 7 feet 9 inches long has four sidewalls covered with tile to a height of 4 feet 6 inches. If an allowance of 19 square feet is made for door and window openings, how many square feet of wall surface are covered with tile? _____

17. A bathroom 6 feet 6 inches by 7 feet 8 inches has a tile floor. Determine the number of square feet covered. (Make no allowance for tub, toilet, etc.) _____

18. The four sidewalls of a room 14 feet 6 inches wide and 17 feet 8 inches long are paneled to a height of 5 feet 6 inches. A deduction of 27 square feet is made for door and window openings in the room. How many square feet of wall surface are covered? _____

19. Determine the number of square feet of surface in a concrete sidewalk 4 feet 10 inches wide and 65 feet 6 inches long. _____

20. How many square feet of cedar are required to line the sidewalls and ceiling of a closet 6 feet wide, 4½ feet deep, and 8 feet high? (Allow 20 square feet for the door.)

21. A lawn is planned on the site illustrated. Deduct the area of the house and drive to find the area in square yards that is to be seeded.

22. At a rate of $2.65 pr square foot, what is the cost of the laminated plastic needed to cover the exposed surface of the illustrated countertop and backsplash? (Ends are not to be covered.)

Unit 22 SURFACE MEASUREMENT— TRIANGLES

BASIC PRINCIPLES OF SURFACE MEASUREMENT

The procedure for finding the area of a rectangle is to multiply the length by the width, $l \times w = A$. The length could be considered as the base and the width considered as the altitude or height, $b \times h = A$.

Example:

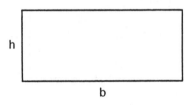

The procedure to find the area of a quadrilateral with only one pair of parallel sides is very similar. Since the parallel sides are not necessarily equal, you must use their average for the base, $\frac{1}{2}(b_1 + b_2) \times h = A$.

Example:

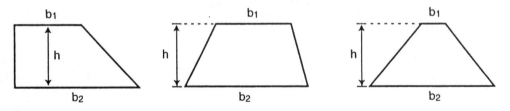

The procedure for finding the area of a triangle is very similar. Think of triangles as having four sides, the fourth being the vertex, which has zero length, $\frac{1}{2}(b + 0) \times h = A$ or $\frac{1}{2}b \times h = A$.

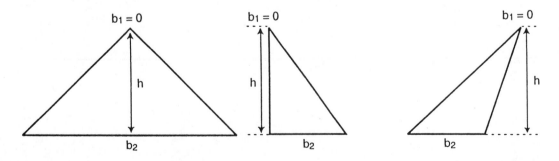

PRACTICAL PROBLEMS

Find the area of each of the following triangles:

1. $h = 16"$ $b = 22"$ _____

2. How many square feet of gable are there in the triangular figure? _____

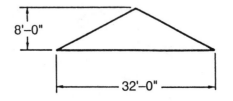

3. A gable has a span of 50 feet and a rise of 10 feet. What is the area in square feet? _____

4. A gable end has a 7-foot 6-inch rise and a span of 30 feet 6 inches. What is its area? _____

5. What is the area of the hip roof shown if the length of one side is 22 feet and the common rafter length is 14 feet 6 inches? _____

Note: A trapezoid is a quadrilateral with only one pair of parallel sides.

$$Area\ of\ a\ trapezoid = \frac{base_1 + base_2}{2} \times height \text{ or } A = \tfrac{1}{2}(b_1 + b_2) \times h$$

6. The building illustrated is 24 feet wide and 36 feet long. The common rafters are 17 feet long with no tails. The length of the ridge is 12 feet. Find the area of the hip roof. (Hint: Consider the roof as two triangles and two trapezoids, each with altitudes equal to the length of the common rafters.) _____

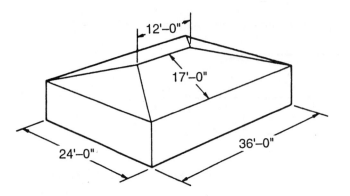

7. A gable roof has a span of 28 feet 6 inches. The height of the ridge above the plate is 14 feet 3 inches. How many square feet of surface are there in one of the gable ends? _____

8. Find the number of square feet of form work required for the end of the concrete retaining wall illustrated.

Note: Use this illustration for problems 9–12.

9. What is the area of a gable that has a rise of 6 feet and a span of 18 feet? _____

10. A building is 30 feet wide and the ridge is 10 feet above the plate line. What is the area of the gable? _____

11. How many square feet of wall space are there in the two gable ends of a house if the ridge is 10 feet above the plate line and the building is 36 feet wide? _____

12. A gable roof has a span of 11 feet 6 inches and a rise of 7 feet 6 inches. What is the total area of the ends of the two gables? _____

13. What is the area of the square hip roof shown if the common rafter length is 16 feet 4 inches? _____

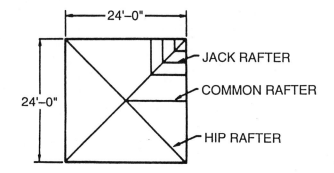

14. A building lot is in the shape of a triangle with a base of 80 feet and an altitude of 50 feet. What is its area in square feet? _____

15. a. Find the area of the shaded portions of the hip roof shown if the length of the common rafter is 12 feet 6 inches. a. _____

 b. What is the area of the entire roof surface shown? b. _____

Unit 23 THE FRAMING SQUARE

BASIC PRINCIPLES OF THE FRAMING SQUARE

The *framing square*, or *steel square*, is made up of two arms: the *blade* (or *body*) and the *tongue*.

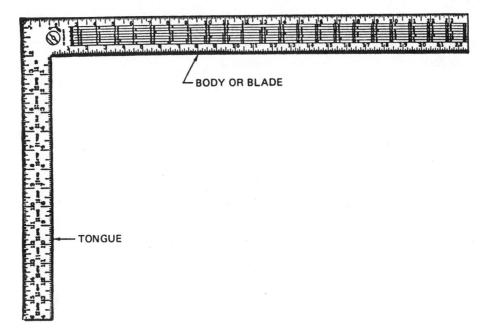

BODY OR BLADE

TONGUE

The blade is generally 24 inches long and 2 inches wide. The smaller arm, called the tongue, is normally 16 inches long and 1½ inches wide. The carpenter's ability to work with decimals and change them to fractions is very important when using a framing square.

Framing squares contain both scales and tables. Manufacturers of squares show a variety of tables, but a carpenter should own one with *rafter* tables.

LENGTH	COMMON	RAFTERS	PER FOOT	RUN	21 63	20 81	20		19 21	18 44	17 69
"	HIP OR	VALLEY	"	"	24 74	24 02	23 32	22 65	22		21 3
DIFF	IN LENGTH	OF JACKS	16 INCHES	CENTERS	28 7/8	27 3/4	26 11/16	25 5/8	24 9/16	23 9/16	
"	"	"	2 FEET	"	43 1/4	41 5/8	40		38 7/16	36 7/8	35 3/8
SIDE	CUT	OF	JACKO	USE	6 11/16	6 15/16	7 3/16	7 1/2	7 13/16	8 1/8	
"	"	HIP OR	VALLEY	"	8 1/4	8 1/2	8 3/4	9 1/16	9 3/8	9 5/8	

Rafter tables are used in determining the length of *common*, *hip*, and *valley* rafters for roofs of various pitches. They are used in determining the difference in lengths of jack rafters and also to aid in making side cuts on hip, valley, and jack rafters. The problems in this unit are limited to the length of common rafters.

Note: Use this illustration for problems 1–6.

FACE OF
RAFTER TABLE

Study the portion of the framing square shown. Under the 8-inch mark, the first row of numerals reads 14 42. This indicates that for each foot of run with a rise of 8 inches, the length of a common rafter is 14.42 inches.

Example: Find the length of a common rafter (without allowing for tail or overhang) in the building shown.

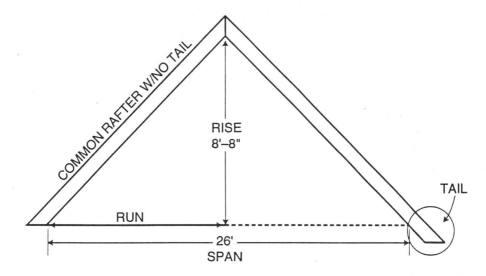

The span (26 feet) divided by 2 equals the run (13 feet). The rise (8 feet 8 inches or 104 inches) divided by the run (13 feet) equals the *rise per foot of run* (8 inches). Thus, the length of common rafters with a rise of 8 inches per foot of run is 14.42 inches per foot of run. The total length of this rafter is 13 times 14.42 inches or 187.46 inches. To convert this measurement to feet and inches, perform the following operations:

$$187.46 \text{ in} \div 12 = 15.6216 \text{ ft}$$
$$15.6216 \text{ ft} = 15 \text{ ft} + 0.6216 \times 12 \text{ in}$$
$$0.6216 \times 12 \text{ in} = 7.459 \text{ in}$$
$$7.459 \text{ in} = 7 + (0.459 \times {}^{16}\!/_{16} \text{ in}) \text{ or } 7{}^{7}\!/_{16} \text{ in}$$

Thus, 187.46 inches equals 15 feet 7⁷⁄₁₆ inches, the length of the common rafters in the illustration.

PRACTICAL PROBLEMS

Using the rafter table, find the length of common rafters in the following problems. Express answers in feet and inches, correct to the nearest ⅛ inch.

1. Run is 7 feet and rise is 10 inches per foot of run. _____

2. Span is 20 feet and rise is 12 inches per foot of run. _____

3. Span is 28 feet and rise is 6 inches per foot of run. _____

FACE OF
RAFTER TABLE

The *pitch* of a roof is the ratio of its rise to its span:

$$\text{pitch} = \frac{\text{rise}}{\text{span}}$$

where rise and span have the same units. Thus, a roof with a ⅙ pitch has a rise of 1 foot per 6 feet of span. This is the same as 12 inches rise per 3 feet (6 ÷ 2) of run, or 4 inches rise per foot of run. To convert quickly from pitch to rise per foot of run, simply multiply pitch by 24:

$$\frac{1}{6} \times 24 = \frac{4 \text{ in rise}}{\text{ft run}}$$

Note: Use this illustration for problems 4–6.

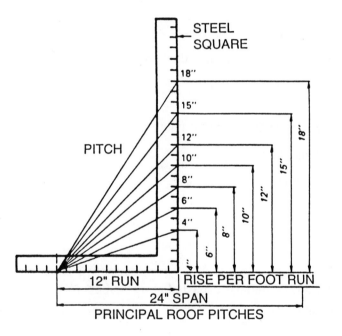

PRINCIPAL ROOF PITCHES

4. Express ⅓ pitch in terms of rise per foot of run. _____

5. Express ½ pitch in terms of rise per foot of run. _____

6. Find the length of a common rafter with ¼ pitch and an 8-foot run. _____

Unit 24 SURFACE MEASUREMENT—
IRREGULAR FIGURES

BASIC PRINCIPLES OF SURFACE MEASUREMENT

There are two fundamental approaches to be used in solving surface measurements of irregular figures. After carefully studying the following methods, select the method best suited to the particular problem you are solving.

Method I

a. Divide the figure into smaller sections.

b. Find the area of each of the smaller sections.

c. Add the areas of the sections to obtain the total area.

Method II

a. Extend the lines to form a simple geometric figure.

b. Find the area of the larger figure.

c. Find the areas of the extra sections.

d. Subtract the areas in step c from the larger area in step b.

PRACTICAL PROBLEMS

1. How many square feet of floor surface are there inside the walls of the plan shown? Use method I to find the total area. _____

2. Find the area of the plan shown by using method II. _____

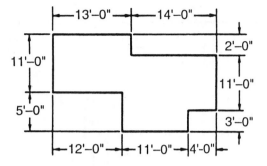

Note: Choice of method for the following problems is left to the student.

3. The plan shown has walls 6 inches thick. A 24-inch border of light-colored carpet is installed around the entire floor.

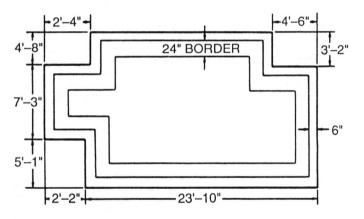

a. Determine the area included within the walls. a. _____

b. Find the area of the 24-inch light-colored carpet. b. _____

Note: Use this illustration for problems 4–6.

4. How many square feet of floor area are there in room A of the illustration? _____

5. Determine the floor area to be covered in room B. _____

6. How many square feet of border surface are there in both rooms if the border is 16 inches wide? _____

7. How many square inches of surface are there in the end of the form for the retaining wall shown? _____

Note: Use this illustration for problems 8–11.

8. A retaining wall, as shown, is 40 feet long. How many square feet of form work are required for the outside finished face, from top to bottom? _____

9. What is the area of the end of the retaining wall above the lower grade? _____

10. Form work is required for the top of the 6-inch steps. How many square feet of form work does it take for the fill side of the retaining wall if it is 40 feet long? _____

11. What is the area of the end of the retaining wall below the lower grade line? _____

Note: Use this illustration for problems 12–14.

12. In room A of the illustration, how many square yards of drywall to the nearest tenth does it take to cover the ceiling? _____

13. How many square yards of drywall to the nearest tenth does it take to cover the ceilings of rooms B and C? _____

14. The walls shown in this figure are 8 feet 6 inches high. How many square yards of drywall to the nearest tenth are required for the walls if the openings total 112 square feet? _____

Unit 25　SURFACE MEASUREMENT— CIRCLES

BASIC PRINCIPLES OF SURFACE MEASUREMENT

The area of circular figures is found using the formula $A = \pi r^2$ (r is the length of the radius of the circle and π is a constant value approximately $^{22}\!/_{7}$ or 3.1416). If remembering this formula is difficult, think of it as πrr (pi railroad). If the diameter of the circle is given instead of the radius, remember that the radius is always one-half the diameter ($^{d}\!/_{2} = r$).

Example: How many square feet of surface area are there in a circular figure with a radius of 8 feet 6 inches?

$Area = \pi r^2$

$Area = 3.1416 \times 8.5 \text{ ft} \times 8.5 \text{ ft} = 226.98 \text{ sq ft}$

PRACTICAL PROBLEMS

Using $\pi = 3.1416$, find the area of each of the following circles:

Express answers for problems 1 and 2 correct to the nearest square inch.

Express answers for problems 3 and 4 correct to the third decimal place.

1. Diameter = 3' 0"　＿＿＿＿＿＿

2. Radius = 6' 0"　＿＿＿＿＿＿

3. Diameter = 8 m　＿＿＿＿＿＿

4. Radius = 7 mm　＿＿＿＿＿＿

5. Cross-sections of pipes A and B are shown. What is the difference to the nearest square inch between their inside areas?　＿＿＿＿＿＿

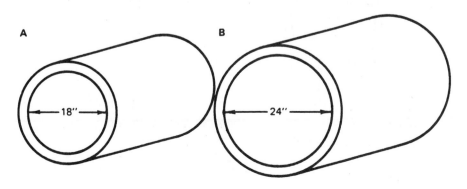

Note: Express the answers correct to the nearest thousandth.

6. Determine the floor area in the semicircular bay window shown. _____

3'-6"

7. How many square feet of ground area are occupied by the curved runway shown? _____

¼ CIRCLE

12'-0"

17'-0"

8. How many square yards of ceiling surface are there in a bay window ceiling if the bay is one quarter of a circle in plan and has a radius of 7 feet 6 inches? _____

9. How many square feet of concrete surface are there in the floor of the silo shown? _____

15'-6"

10. A circular plywood table top with a diameter of 36 inches is covered with laminated plastic. The cover piece is cut from a sheet 36 inches square. How many square inches of material are wasted? _____

11. A stencil is needed to aid in spray painting a playground. If the stencil is cut from a 4-foot by 4-foot piece of plywood, how much waste is created? _____

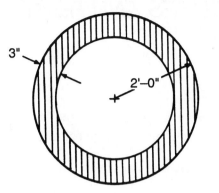

3"

2'–0"

12. A rough floor is laid under the space occupied by the gymnasium track shown. Determine the area occupied by the track. Do not include the space inside the track. _____

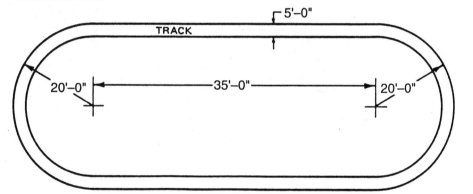

5'–0"

TRACK

20'–0" 35'–0" 20'–0"

13. Find the cross-sectional area of the wooden block illustrated. _____

3"

2"

5"

Unit 26 VOLUME MEASUREMENT—
CUBES AND RECTANGULAR
SOLIDS

BASIC PRINCIPLES

Volume is measured in *cubic* inches, *cubic* feet, *cubic* yards, and so forth. It is never measured in inches, feet, yards, etc. One inch times one inch times one inch equals one *cubic* inch.

To find the volume of a rectangular solid, first express all dimensions in the same unit of measurement, then multiply width by length by height.

Example: Find the volume of a rectangular solid that is 2 feet wide, 1 yard long, and 18 inches high.

First change the problem to read 2 feet wide, 3 feet long, and 1½ feet high. The answer will be in cubic feet.

$V = w \times l \times h$

$V = 2' \times 3' \times 1\frac{1}{2}'$

$V = 9$ cu ft

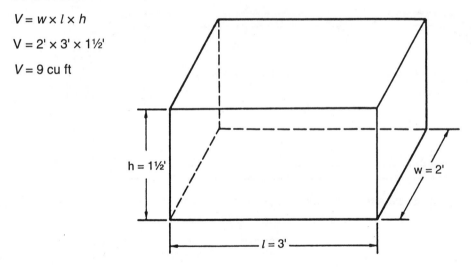

PRACTICAL PROBLEMS

Find the volume of the following cubes:

1. Side = 9' _____

2. Side = 5 cm _____

3. Side = 8' 11" _____

Find the volume of each of the following rectangular solids:

4. Length = 6" Width = 5" Height = 4" _____

5. Length = 8 m Width = 5 m Height = 3 m _____

6. Length = 10' Width = 8' Height = 2½' _____

7. Length = 3' Width = 2' Height = 6" _____

8. How many more cubic feet are there in container A than in container B? _____

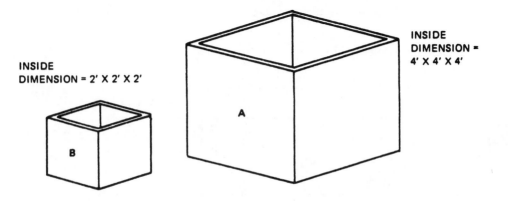

INSIDE
DIMENSION = 2' X 2' X 2'

INSIDE
DIMENSION =
4' X 4' X 4'

A

B

9. How many cubic feet of space are there in a storage area 8 feet wide, 16 feet long, and 7 feet high? _____

10. How many cubic yards of concrete are needed to pour eight footings, 20 inches by 20 inches by 8 inches? _____

LALLY
COLUMN

20"

8"

20"

11. How many cubic yards of earth must be removed for the excavation shown
 if the depth is 8 feet? _____

 Note: Excavations are usually computed in terms of cubic yards.

Note: Use this illustration for problems 12–14.

CROSS SECTION OF SIDEWALK

12. The illustration shows a cross section of a concrete sidewalk. How many
 cubic yards of earth does it displace if the walk is 40 feet long? _____

13. How many cubic yards of concrete are there in the sidewalk shown? _____

14. How many cubic yards of cinders are required for the sidewalk shown? _____

15. A large building requires a basement 9 feet deep, 78 feet wide, and 96 feet
 long. How many cubic yards of earth must be removed if an additional 15%
 must be removed for the foundation and floor? _____

16. A storage space in a basement is 8 feet 6 inches by 6 feet 8 inches by 8 feet 8 inches. What is the number of cubic feet of storage space? _____

17. The illustration shows a section of a concrete foundation wall and footing. How many cubic yards of concrete does it contain? _____

18. An apartment house requires a concrete foundation wall 9 inches thick and 3 feet high. The size of the building is 38 feet by 82 feet.

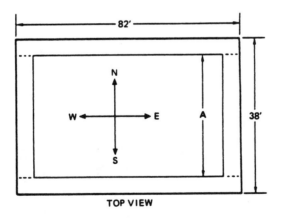

TOP VIEW

 a. How many cubic feet of concrete are needed for both the north and south walls? a. _____

 b. Find missing dimension A. b. _____

c. How many cubic feet of concrete are needed for the east and west
walls? c. _____

d. How many cubic feet of concrete are there in the entire foundation? d. _____

19. A building is 28 feet wide and 44 feet long. The excavation for the footing is
18 inches wide and 6 inches deep. How many cubic feet of earth must be
removed in digging a trench for just the footing? Assume the excavation for
the foundation to have been previously dug. _____

Unit 27 BOARD MEASURE

BASIC PRINCIPLES OF BOARD MEASURE

A *board foot* (bd ft) is defined as the equivalent of a piece of wood measuring one foot wide, one foot long and one inch thick. In each of the sketches, the number of board feet is shown.

| 1 BOARD FOOT | 4 BOARD FEET | 6 BOARD FEET |

To calculate the board measure in any quantity or piece of lumber, use the formula:

$$\text{Board feet} = T \times W \times L$$

in which T = thickness (expressed inches), W = width (expressed in feet), and L = length (expressed in feet).

Note: Board measure problems are figured using standard dimensions for thickness and width. Lumber less than 1" thick is counted as a full inch. Standard *thicknesses* are 1", 5/4", 6/4", 2", 3", 4", 6", and so on. Standard *widths* are 2", 3", 4", 5", 6", 8", 10", and 12". For other widths or thicknesses, count the next standard dimension above the exact dimension desired.

Example: To find the number of board feet in the piece of lumber shown, use $T = 1$ inch, $W = 6/12$ foot, and $L = 4$ feet.

$$\text{Board feet} = T \times W \times L = 1 \times \tfrac{6}{12} \times 4 = 2 \text{ bd ft}$$

Notice that the dimension for width must be converted to feet, while thickness remains in inches.

In the board feet problems in this text, rough stock dimensions are used. *Rough stock* is lumber that is not planed or surfaced. *Dressed stock* is lumber that has been surfaced. In dressing a board, the width and thickness can be reduced by ½ inch. For example, a 2 × 4 will measure about 1½ inches by 3½ inches when dressed.

PRACTICAL PROBLEMS

Find the number of board feet in each of the following quantities:

1. 5 pieces of 1" × 6" × 18' _____

2. 34 pieces of 2" × 4" × 16' _____

3. 62 pieces of 1" × 10" × 18' _____

Find the number of board feet in the following quantities:

4. 18 pieces of ½" × 4" × 16' _____

5. 8 boards, 1½" thick, 22" wide, 16' long _____

6. 25 pieces, ¾" × 3" × 12' _____

7. Determine the total number of board feet of select sugar pine in the following list:

 10 pieces of ⁵⁄₄" × 4" × 16' _____
 6 pieces of ⁵⁄₄" × 10" × 18' _____
 1 piece of ⁵⁄₄" × 12" × 12' _____

Note: Use this illustration for problems 8–13.

8. The building shown requires two 4 × 6-inch sills, each 38 feet long. They are constructed from four pieces of 4 × 6-inch timber, each 20 feet long. This allows for a splice 2 feet long in the center of each sill. How many board feet of lumber must be ordered? _____

9. Determine the board feet of stock to be ordered for each floor joist running across the 13-foot span. (Floor joists are sold in multiples of 2 feet in length.) _____

10. How many board feet of stock are to be ordered for all of the floor joists for the 13-foot span? (To find the number of joists required, divide the distance by the spacing, 16 inches o.c., and add 1 joist for a starter.) _____

11. Find the number of board feet that must be ordered for each floor joist running across the 15-foot span. _____

12. Find the number of board feet of subflooring to be ordered for the entire floor surface if 1 × 6-inch matched boards are to be used. (The number of board feet to be ordered is found by determining the area and adding 20% for matching and waste.) _____

13. How many board feet of 1 × 3-inch oak flooring must be ordered for the entire floor, as shown, if an allowance of 38% is made for waste and matching? _____

Unit 28 VOLUME MEASUREMENT— CYLINDERS

BASIC PRINCIPLES OF VOLUME MEASUREMENT

The procedure for finding the volume of a cylinder calls for the use of another new formula: $V = \pi r^2 h$, where $r =$ radius of the circular end, $\pi = 3.1416$, and $h =$ the height of the cylinder when viewed with the circular surface as the base. All measurements must be in the same units before proceeding to multiply.

Example: Find the volume of a circular cylinder with a radius of 12 feet and height of 30 feet.

$V = \pi r^2 h$
$V = 3.1416 \times 12 \text{ ft} \times 12 \text{ ft} \times 30 \text{ ft}$
$V = 3.1416 \times 144 \text{ sq ft} \times 30 \text{ ft}$
$V = 3.1416 \times 4,320 \text{ cu ft}$
$V = 13,571.712 \text{ cu ft}$

PRACTICAL PROBLEMS

Note: Let $\pi = 3.1416$. Express all answers correct to the nearest hundredth.

1. A cross section of a cylindrical cistern is shown. What is the volume of the cistern?

4'–0"

2'–0"
DIAMETER

2. What is the capacity in gallons of a wooden storage tank that has an inside diameter of 16 feet and an inside height of 18 feet 9 inches? One cubic foot contains 7.481 gallons.

Note: Use this illustration for problems 3 and 4.

3. What is the capacity in cubic feet of the circular silo shown?

4. How many cubic yards of concrete are needed for the wall of this silo?

5. A loading dock is supported by two solid concrete piers. The piers are each 2 feet in diameter and 10 feet high. How many cubic feet of concrete do the two piers contain?

6. Sections of wrought iron pipe, having an inside diameter of 4 inches, are filled with concrete and used for posts to support the main floor girder in the basement of a residence. If these posts are 8 feet long, and there are 6 in all, how many cubic feet of concrete are required for the job?

7. How many cubic yards of concrete are needed for a foundation that is 14 feet 6 inches in diameter and 2 feet 6 inches deep? _____

Note: Use this illustration for problems 8–12.

8. The posts in the illustration are set in concrete. All posts and concrete casings are cylindrical. How many cubic feet of earth are excavated for setting both identical posts A and D? _____

9. How many cubic feet of concrete are needed to pour around both posts A and D? _____

10. The two posts B and C are alike. How many cubic feet of earth are removed for setting both posts B and C? _____

11. How many cubic feet of concrete are needed to pour around post B? _____

12. Determine the number of cubic feet of concrete needed to set the four posts shown if posts A and D are changed to 8 inches in diameter, and posts B and C are changed to 12 inches in diameter. All other dimensions remain unchanged. _____

13. What is the volume in cubic inches of the pattern block shown? _____

Unit 29 WEIGHT MEASURE

BASIC PRINCIPLES OF WEIGHT MEASURE

It is important for carpenters to solve problems involving weight. Loads must be kept within safe bearing capacities of various soils. When hauling materials, the safe capacities of the truck must not be exceeded. When designing a structure, the live load and dead load must be figured to use beams of the most efficient size. This unit contains problems carpenters will encounter in their daily work.

In the English system, common units of weight and their equivalents are ounces, pounds, and tons.

The basic metric unit of weight is a gram.
A gram has about the weight of a paper clip.

Kilogram: 1,000 grams; a little more than 2 pounds (about 2.2 pounds)
A pound weighs slightly less than ½ kilogram.
1 pound = 0.45 kilogram

1 KILOGRAM 1 POUND

ENGLISH UNITS OF WEIGHT		
Weight Measure	Symbol	Equivalent
ounce	oz	standard unit of weight
pound	lb	16 ounces
ton	T	2,000 pounds

METRIC UNITS OF WEIGHT		
Weight Measure	Symbol	Relation to Gram
gram	g	standard unit of weight
kilogram	kg	1,000 grams
metric ton	t	1,000,000 grams

PRACTICAL PROBLEMS

1. If asphalt shingles weigh 235 pounds per square (100 square feet), what is the weight of the shingles needed to cover the area shown? _____

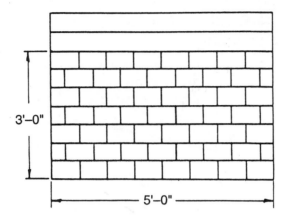

2. The safe bearing capacity of a soil is 2,500 pounds per square foot. What area of footing is required to sustain a load of 60,000 pounds? _____

3. The total weight of a tank when completely filled is 35,000 pounds. How much weight per square foot is carried by a footing 22 square feet in area? _____

4. If a live load of 70 pounds per square foot is added to the dead load, what is the live load on a floor measuring 40 feet by 20 feet? _____

5. The type of concrete generally used weighs about 145 pounds per cubic foot. Estimate the weight of a concrete wall 20 feet long, 6 feet high, and 14 inches thick. _____

6. I beams are ordered and sold by the size and weight per linear foot. At 40 pounds per foot, how much does a 10 × 4¾-inch I beam, 24 feet long, weigh? _____

7. The weight of a certain quality of sheet steel is 487.7 pounds per cubic foot. Find the weight of a plate of this steel that is ⅛ inch thick and has 24 square feet of surface on each face. _____

8. Water weights 1,000 kilograms per cubic meter. Find the weight of the water contained in the holding tank shown if the tank is filled to the level indicated. _____

DEPTH OF LIQUID = 0.5 m
INSIDE WIDTH = 1.5 m
INSIDE LENGTH = 2 m

9. Sheet copper is ordered by size and weight per foot, but is sold only by the actual weight of the material. A carpenter orders a sheet of copper weighing 16 ounces per square foot. The sheet is 16 inches by 16 feet. How many pounds does the sheet weigh? _____

10. A contractor purchases a half ton of 8d common nails for $620. What is the unit price per box (50 pounds)? _____

11. In house framing, 20 pounds of 8d common wire nails are estimated for each 1,000 square feet of sheathing or subflooring. How many pounds of nails should be ordered for 12,000 square feet of sheathing? _____

12. Sand is estimated by the cubic yard, but in certain localities it is sold by the ton. Find the weight of 5 cubic yards of washed sand, allowing 95 pounds per cubic foot of material. _____

13. When wood shingles are laid 4½ inches to the weather, an allowance of 5¾ pounds of 4d nails is made for each square of shingles. How many pounds of shingle nails must be ordered to lay 22½ squares of shingles at this exposure? _____

14. At $7.25 per ton, find the cost of 15 cubic yards of washed and screened gravel weighing 106 pounds per cubic foot. _____

15. How many tons of washed gravel weighing 105 pounds per cubic foot will a truck carry if the inside body measurements are 5 feet 6 inches wide, 14 feet long, and 2 feet 6 inches high? (The material is leveled off to the top of the body.)

16. Determine the total weight of the concrete wall and footing illustrated if the concrete used weighs 145 pounds per cubic foot.

17. If roll roofing weighs 90 pounds per roll and covers an area of one square, find the weight of roll roofing used to cover both sides of a gable roof, each side measuring 42 feet long and common rafters 16 feet long.

18. Find the weight to the nearest hundredth of the steel angle bracket shown if the material weighs 0.28 pound per cubic inch.

19. What is the total weight of 15 pieces of bar stock if each piece has dimensions as indicated? _____

 Note: Stock weighs 30 pounds per linear foot.

Powers and Roots

Unit 30 APPLICATIONS OF EXPONENTS IN FORMULAS

BASIC PRINCIPLES OF POWERS

An exponent is a small figure written to the right and slightly above a given quantity. Exponents are used as a convenient way to indicate the number of times a quantity is to be multiplied by itself.

Example: $5^2 = 5 \times 5 = 25$
$10^3 = 10 \times 10 \times 10 = 1,000$

Formulas used in this unit involve the use of exponents.

Examples: Area of a square $A = s^2$
Volume of a cube $V = s^3$
Area of a circle $A = \pi r^2$
Volume of a cylinder $V = \pi r^2 h$

PRACTICAL PROBLEMS

1. Find the area of the 8-inch square shown.
 $(A = s^2)$

 8″

 8″

Find the area of each of the following squares:

2. Side = 12″ _____

3. Side = 7 mm _____

4. Find the volume of the 8-cm cube shown.
 $(V = s^3 \text{ or } V = s \times s \times s)$

 8 cm

 8 cm

 8 cm

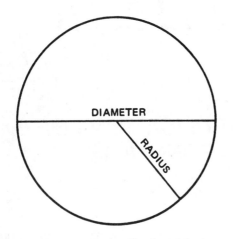

5. Find the area of a circle with a radius of 2 feet 7 inches. _____

6. Find the area of a circle 2 meters in diameter. _____

7. Find the volume of a cylinder
 with a radius of 8 inches and
 a height of 14 inches.
 ($V = \pi \times r^2 \times h$) _____

8. Find the volume of a cylinder
 with a diameter of 16 inches
 and a height of 2 feet. _____

9. A contractor makes columns by filling 20 lengths of pipe, 4 inches in
 diameter by 8 feet long, with concrete. How many cubic feet of concrete are
 used? _____

10. Find the area of the rectangle shown. _____

11. What is the area of a floor 16 feet square? _____

12. Find the area of a floor 14 feet square. _____

13. What is the ceiling area of a room that measures 12 feet on each side? _____

14. What is the area of a circular floor that has a diameter of 16 feet? _____

15. Find the volume of concrete required for the silo base in the figure below. _____

(Foundation is 1½ feet deep with outside diameter of 20 feet and inside diameter of 18 feet.)

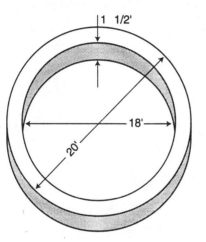

16. A circular tower has a radius of 12 feet. What is the floor area? _____

17. A heavy machine requires a circular concrete base that is 13 feet in diameter and 10 inches deep. Find the volume of the base. _____

18. How many cubic feet are contained in a footing 2 × 2 × 2 feet? _____

 # Unit 31 USING SQUARE ROOT TO FIND SIDES OF RIGHT TRIANGLES

BASIC PRINCIPLES OF SQUARES AND ROOTS

The *square* of a number is the result of multiplying a number by itself.

Examples: The *square* of 3 is 9 $3^2 = 3 \times 3 = 9$

The *square* of ¼ is ¹⁄₁₆ $\frac{1}{4}^2 = \frac{1}{4} \times \frac{1}{4} = \frac{1}{16}$

The *square root* is the number that, when multiplied by itself, produces the given number. The symbol for square root is $\sqrt{\ }$.

Examples: The *square root* of 9 is 3 $\sqrt{9} = 3$

The *square root* of ¹⁄₁₆ is ¼ $\sqrt{\frac{1}{16}} = \frac{1}{4}$

Square roots can be found by using a table of powers and roots, as found in the Appendix (Table IV). Using a pocket calculator is also very practical.

Caution: When using a calculator, *always* make a mental estimate first.

The *hypotenuse* of a right triangle is the side opposite the right angle. The remaining two sides are called the *legs* of the triangle. When the lengths of any two sides of a right triangle are known, the length of the third side can be found by using a rule called the Pythagorean theorem.

The Pythagorean Theorem: The square of the hypotenuse of a right triangle is equal to the sum of the squares of the two legs.

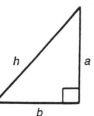

Stated as a formula $h^2 = a^2 + b^2$ where h = hypotenuse, a = altitude, and b = base.

This theorem can be illustrated by the areas of the squares in the figure:

$$h^2 = a^2 + b^2$$
$$25 = 16 + 9$$

It is also true that:

$$a^2 = h^2 - b^2$$
$$16 = 25 - 9$$

and

$$b^2 = h^2 - a^2$$
$$9 = 25 - 16$$

h = HYPOTENUSE
a = ALTITUDE
b = BASE

The following are equivalent forms of this theorem:

$$h = \sqrt{a^2 + b^2} \qquad a = \sqrt{h^2 - b^2} \qquad b = \sqrt{h^2 - a^2}$$

To find one side of a right triangle when two sides are known:

1. Choose the appropriate form of the formula.
2. Substitute the lengths of the two known sides.
3. Perform the operations.

Example: If h = 13 in and b = 12 in, find a.

1. $a = \sqrt{h^2 - b^2}$
2. $a = \sqrt{(13 \text{ in})^2 - (12 \text{ in})^2}$
3. $a = \sqrt{169 \text{ in}^2 - 144 \text{ in}^2} = \sqrt{25 \text{ in}^2} = 5$ in

Study the Framing Square method for finding the length of one side of a right triangle.

The Framing Square Method: Carpenters often use the framing square to approximate the length of one side of a right triangle when the other two sides are known. The scales on the outer edges of the blade and tongue are divided into twelfths on the back side of the square. Using the back of the square, a measurement given in feet and inches can be located by reading the twelfths as inches and the inches as feet. If measurements are given in inches and fractions of an inch, the face of the square should be used.

To find the hypotenuse when the lengths of the two legs are known:

1. Locate the length of one leg on the outer edge of the blade.

2. Locate the length of the second leg on the outer edge of the tongue.

3. The distance between these two locations, measured by a carpenter's rule or a second square, is the length of the hypotenuse.

Example: The two legs of a right triangle measure 9 feet 7 inches and 13 feet 10 inches. Find the length of the hypotenuse.

1. On the back of the square, locate 13¹⁰⁄₁₂ inches on the body. This will represent 13 feet 10 inches.

2. On the tongue, find 9⁷⁄₁₂ inches. This will represent 9 feet 7 inches.

3. The distance between these two points read on a carpenter's rule is slightly greater than 16¾ inches. This is interpreted as slightly greater than 16 feet 9 inches.

To find the leg when the lengths of the hypotenuse and the other leg are known:

1. Locate the known leg on the outer edge of the tongue.

2. Locate the length of the hypotenuse on the body of a second square or a carpenter's rule.

3. Place them as shown in the following figure and example. Then, read the length of the unknown side on the body of the second square or carpenter's rule.

Example: Find the missing leg of the triangle with a hypotenuse of 13 inches and a leg of 5 inches.

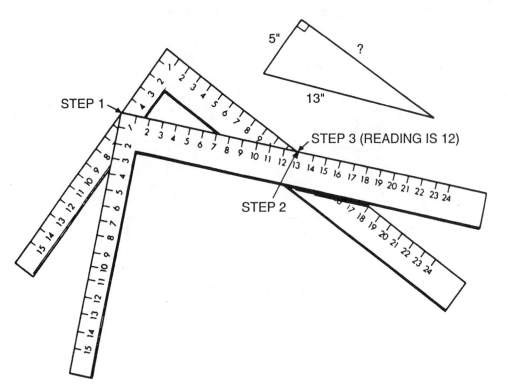

1. Locate 5 inches on the tongue of the first square and place the heel of a second square at that point.

2. Locate a hypotenuse of 13 inches on the body of the second square.

3. At the point where the 13 crosses the first square, read the length, 12 inches, on the body of the first square.

PRACTICAL PROBLEMS

Unless otherwise indicated, find square roots of all problems to the nearest hundredth. Find the following square roots:

1. $\sqrt{81}$ _____

2. $\sqrt{100}$ _____

3. $\sqrt{169}$ _____

4. $\sqrt{361}$ _____

5. $\sqrt{529}$ _____

6. $\sqrt{1,024}$ _____

Note: Use the illustration of the wall brace shown for problem 7. Find the length of the brace correct to the nearest eighth inch.

7. Wall height = 6', run of brace = 9' _____

Note: Use the illustration of a stairway for problems 8–10. In each problem, find the length of the stringer required for the stairway.

8. Total rise = 12' 0", total run = 16' 0" _____

9. Total rise = 5' 2", total run = 7' 6" _____

10. Total rise = 2' 7", total run = 3' 9" _____

Note: Use this illustration for problems 11–15.

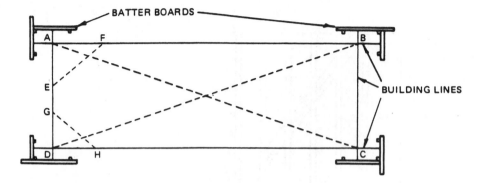

In the illustrated figure, rectangle ABCD represents the lines of excavation for the foundation of a house.

11. If line AB = 45 feet and AD = 27 feet, what is the length of diagonals AC and BD? _____

12. Find the lengths of the diagonals AC and BD if the side AB is 43 feet 6 inches and AD is 32 feet 6 inches. _____

13. In checking to find out if the corner D is square, GD is laid off equal to 9 feet and DH is laid off equal to 12 feet. What should the length of the GH be? _____

14. What are the lengths of diagonals AC and BD if AB and BC are each 41 feet 6 inches? _____

15. To square up the corner A, AE is laid off equal to 6 feet and AF equal to 8 feet. What should the length of line EF be? _____

Note: Use this illustration for problems 16–21.

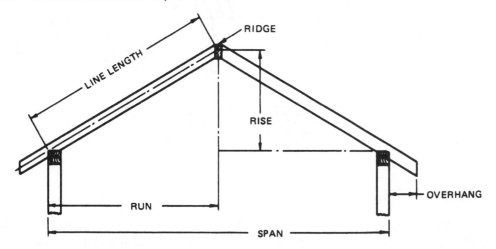

Note: *Span*, *run*, and *overhang* are terms used by carpenters for horizontal dimensions. Study the illustration and become familiar with these terms before proceeding.

For problems 16–18, find the length of the common rafters to the nearest inch. (For these problems, do not allow for the ridge board.)

16. Span = 28', total rise = 7', overhang = 1' 6" _____

17. Span = 14', total rise = 4' 8", overhang = 1' 3" _____

18. Span = 27', total rise = 9', overhang = 2' 8" _____

19. In the illustration of the roof, what is the line length of the rafter if the run is 12 feet and the rise is 8 feet? _____

20 A common gable roof, similar to the one shown, has a span of 26 feet and a rise of 5 feet. What is the line length of the common rafter? _____

21. Determine the line length of a common rafter for a roof that has a run of 16 feet and a rise of 10 feet 8 inches. _____

Estimating

Unit 32 GIRDERS, SILLS, AND FLOOR JOISTS

BASIC PRINCIPLES OF GIRDERS

Girders are the large beams supporting the first floor of a house. These carry the ends of the floor joists where there is no wall. Girders are usually supported at the ends by the outside foundation walls, and between the walls by *columns* or *posts*.

Columns or posts are spaced at intervals that are determined by the ability of the beam to support the load it is designed to carry. The size of the girders and the spacing of the columns are determined by the architect and are shown on the working drawing of the foundation plan. Joints in girders should be made over piers or columns. A four-inch bearing surface under each end of a girder is usually considered a minimum.

In listing solid girders, allowance must be made for the joists and bearing as shown in the diagrams. A *built-up girder* is made of several pieces of lumber nailed together, as in the illustration.

PRACTICAL PROBLEMS

Note: An estimate for the length of stock required for built-up girders must include waste. Girder stock may be purchased only in multiples of 2 feet.

1. The girder span in a house is 29 feet 0 inches. How many pieces of 2 inch by 10 inch by 10 foot lumber are needed to build up a 6-inch by 10-inch girder?　　　　　　　　　　　　　　_____

Note: Lumber is measured in board feet. A board foot is the equivalent of a piece of wood measuring 1 square foot, 1 inch thick. Material that measures 1 or more whole number of inches plus a fraction of an inch is counted as the next whole inch, unless it is of standard thickness. (See p. 109.)

Examples: 1½ inch width is counted as 2 inches
　　　　　　　⅝ inch thickness is counted as 1 inch

2. A house requires two solid 8 × 12-inch girders running parallel to each other. The span is 31 feet 6 inches. How many board feet are required?　　　_____

3. How many board feet of material are required to construct a 6 × 10-inch built-up girder for a building 30 feet long?　　　　　　　　　　_____

4. From the information given, determine the number of board feet of material required for the solid girder in the illustration.　　　　　　　　_____

5. A warehouse is 156 feet long with a span of 60 feet. The 6 × 8-inch girders
 are spaced 12 feet o.c. (on center). The span equals the length of the girder.

 a. How many girders are needed? a. _____

 b. How many board feet are required? b. _____

6. How many columns are required under a 40-foot girder, supported on each
 end by the foundation, if they are spaced 8 feet o.c.? _____

BASIC PRINCIPLES OF SILLS

A sill rests on top of the foundation and is anchored to it by bolts embedded in the top of the foundation.
The distance of the sill from the outside of foundation is usually equal to the thickness of the wall
sheathing. The sill serves as a base on which the ends of the first-floor joists rest. In some cases, the
wall studs are nailed to the sill. Standard sill stock lengths are multiples of two feet, up to 16 feet long.

HEADER JOIST FLOOR JOIST
SILL
PLATE
2" x 10"
2 "x 8"

BOX SILL
(most commonly used)

PRACTICAL PROBLEMS

7. How many board feet are required for a sill 2 inches by 8 inches and 24 feet
 long? _____

8. How many 14-foot lengths of 2 × 6 inches are needed to construct sills for a
 building 40 feet wide and 100 feet long? _____

9. A 4 × 6-inch sill is installed on a foundation 40 × 32 feet. A built-up sill 4 × 6
 inches is constructed using doubled 2 × 6-inch lumber. How many linear
 feet of stock are needed? _____

BASIC PRINCIPLES OF FLOOR JOISTS

Joists are beams used to support a floor or ceiling. Usually the size, spacing, and direction are included on the floor plans. Floor joists run at right angles to the girder or wall that carries them. For purposes of listing or estimating, they are considered to extend to the outside face of the sills. The number of joists required may be determined by dividing the length of the building by the on center spacing of the joists. To this number add 1 for the starter.

Example: For a building 48 feet long with joists spaced 16 inches o.c., the number of joists needed is found as follows:

16 inches equals ⅓ feet
48 feet divided by ⅓ feet = 36 joists
36 joists + 1 joist = 37 joists

Standard length of joist stock is between 8 feet and 16 feet in multiples of 2 feet.

JOIST LAPPED ON
SOLID GIRDER

GIRDER
WITH LEDGER STRIP

JOIST HUNG
ON SOLID GIRDER

PRACTICAL PROBLEMS

10. A building with a 26 × 40-foot foundation has floor joists 16 inches o.c. as shown. How many board feet of 2-inch by 10-inch by 14-foot joists are needed for the first floor?

JOIST SPACING = 16" ON CENTER

2" × 10" × 14' JOIST

BUILT-UP GIRDER

26'–0"

11. a. How many floor joists are required in a building 32 feet long by 26 feet wide if the joists are spaced 16 inches o.c. and there is one girder in the center of the building?

 a. _____

 b. If 2-inch by 10-inch by 16-foot joists are used, how many board feet are needed for the joists?

 b. _____

12. A house requires 44 pieces of 2-inch by 10-inch by 16-foot lumber for first floor joists as shown.

 a. How many board feet are required?

 a. _____

 b. What is the approximate length of the house if the joists are spaced 16 inches o.c.?

 b. _____

Unit 33 ROUGH FLOORING

BASIC PRINCIPLES OF ROUGH FLOORING

Rough flooring, or *subflooring*, is usually made of plywood laid directly on the floor joists, upon which a finished floor is laid. Subflooring sheets are usually installed at right angles to the floor joists.

SILL
JOISTS

PLYWOOD SUBFLOOR
(4' x 8' SHEETS)

HEADER JOIST

The use of *plywood* for subflooring has become standard, largely because of its ease of installation and the resulting savings in labor costs. Sold by the square foot or by the sheet, the most common sheet size is 4 × 8 feet. The amount of plywood needed may be estimated by the square foot. Plywood should be installed with the grain direction of the outer plies running at right angles to the joist. The sheets should be staggered so that end joists in adjacent panels break over different joists.

For the purpose of practicing mathematics, some of the problems in this unit are based on the use of *board* subflooring. When boards are used for subflooring, allowance must be made for waste. This allowance is usually a percentage of the area to be floored. The percentage to be added depends on the width of the boards and how they are to be laid.

DIAGONAL

RIGHT ANGLE

Usually, for 1 × 6-inch matched boards applied at right angles to the floor joists, 25% of the area to be covered is added for waste and matching. For 1 × 8-inch matched boards laid at right angles to floor joists, 20% is added. Another 5% is added when rough floorboards are to be laid diagonally. If there are any openings in the floor, the area of the opening is deducted from the total area.

PRACTICAL PROBLEMS

1. In the illustration, a platform step is shown. How many board feet of subflooring are needed to cover the platform if 1 × 6-inch matched boards are used? (Add 25% for waste and matching.) _____

2. A two-story building 40 × 60 feet is to have subflooring laid at right angles to the joists. How many board feet of 1 × 8-inch matched boards are needed to cover both floors? (Add 20% for waste and matching.) _____

3. How many square feet of area must be covered on a fountain 28 feet wide and 36 feet long? _____

4. Estimate the number of square feet of plywood needed to completely cover a floor 28 feet wide and 48 feet long. _____

5. Figure the quantity of subflooring needed when the plans for a two-story building show a total 864 square feet for each floor. Deduct 311 square feet for baths, kitchen, and entrance. Add ⅓ of the area for matching. When doing this problem, change all fractions to the next whole number. _____

6. Deadening felt is laid over the subflooring in problem 5. Deadening felt has 500 square feet in a roll. How many rolls must be ordered? _____

Unit 34 WALL PLATES

BASIC PRINCIPLES OF WALL PLATES

Plates are the horizontal structural members to which the tops and bottoms of studs are secured. The plate to which the bases of the studs are nailed is sometimes called the *sole plate*.

Top and bottom plates of walls and partitions are considered separately from the studs. Plates running along the top of the studs on the exterior walls are made up of 2 × 4-inch or 2 × 6-inch lumber nailed together with a lap joint at the corners. These usually act as a seat for the second-floor joists or as a seat for the rafters in a one-story house. (See illustration.)

Bearing partition plates are made the same as the plates of exterior walls. In certain types of construction, the bearing partition studs rest on the subfloor above the girder; therefore, a single 2 × 4-inch member is used as a base to which the studs are fastened. In some cases, the outside walls also rest on top of the subfloor and also require a sole plate.

No allowance is made for openings when estimating the linear footage of plates. Express the linear footage as the number of standard lengths required by dividing the standard length into the total linear footage.

EXTERIOR PLATE WITH CORNER JOINT

PARTITIONS AND WALLS

PRACTICAL PROBLEMS

1. How many board feet of 2 × 6-inch stock are needed to construct the sole
 and top plates shown? _____

2. How many board feet of single sole and double top plate of 2 × 4-inch stock
 are needed for the exterior walls in the plan shown? _____

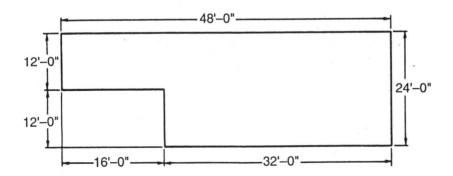

3. How many pieces of 8-foot-long 2 × 4s and how many pieces of 12-foot-long
 2 × 4s are required for double top plates and single sole in the figure below? _____

4. Find the number of 2 × 4-inch 16-foot lengths required for the outside double plates of a house 24 × 42 feet.

5. How many 16-foot lengths are required for the exterior double plates of a house that is 32 × 48 feet?

Unit 35 STUDDING AND FIRE OR DRAFT STOPS

BASIC PRINCIPLES FOR STUDS AND FIRE STOPS

Studs are the vertical members of outside walls or partitions to which sheathing, paneling, or laths are fastened. Studs for small frame buildings are 2 × 4 inches or 2 × 6 inches. When 2 × 6-inch framing is used for exterior walls, the interior partitions are still usually 2 × 4 inches.

A common practice, and a fairly accurate way of estimating the number of studs to be used, is to allow one stud for every linear foot of wall and partition when studs are set 16 inches o.c. When studs are placed 2 feet o.c., allow 3 studs for every 4 feet of wall. This surplus allows for the doubling at corners and around openings. Studs may be obtained in multiples of 2 feet in length, or may be purchased precut to build 8-foot walls.

To estimate the number of studs required in a building, the following steps are usually followed:

1. Calculate the perimeter of a building to determine the number of studs needed for the outside walls.

2. Determine the number of linear feet of all interior partitions.

3. Add the perimeter and the number of linear feet of all partitions.

4. For 2 × 4-inch studs spaced 16 inches o.c., allow one stud for every linear foot of the total. For 2 × 6-inch studs spaced 2 feet o.c., allow 3 studs for every 4 linear feet of the total.

A *fire stop* is material inserted in the space between studs or joists to prevent the vertical spread of fire. It is placed so that it will block the passage of flames or air currents upward or across the building.

PRACTICAL PROBLEMS

1. In the illustration, the fire stops occur between the joists at the second-floor level. How many linear feet of joist material are needed for the stops? _____

2. A house is 24 feet wide and 32 feet long. The studs are spaced 16 inches o.c. At the first-floor level, how many linear feet of fire stops should be used around the outside walls? _____

3. How many outside studs, spaced 24 inches on center, are required to construct the outside walls of a building that is 26 feet wide and 38 feet long? _____

4. How many board feet of 2 × 6-inch stock are required for the studs in the outside walls of a building 26 feet by 38 feet if the studs are spaced 24 inches o.c.? _____

5. How many linear feet of outside studding 8 feet long are required for a house 24 feet by 42 feet? Studs are 16 inches o.c. _____

6. Estimate how many board feet of studding are needed for a house with the following specifications. (All studs are spaced 16 inches o.c.)

 a. Overall measurements are 24 feet wide by 32 feet long. (All studs are 2 inches by 4 inches and 10 feet long.)

a. _____

 b. One partition, 32 feet long; one, 14 feet long. (All studs are 2 inches by 4 inches and 8 feet long.)

b. _____

 c. Three partitions, 10 feet long; one, 14 feet long. (All studs are 2 inches by 4 inches and 8 feet long.)

c. _____

 d. One partition, 6 feet long. (All studs are 2 inches by 4 inches and 8 feet long.)

d. _____

 e. Estimate the total number of studs required for the outside walls and partitions found in parts (a) through (d).

e. _____

7. A house has 120 linear feet of outside walls with studs spaced 24 inches o.c. and 194 partitions with studs spaced 16 inches o.c. Estimate the number of studs that are required for all the studding.

Unit 36 RAFTERS

BASIC PRINCIPLES FOR RAFTERS

Rafters or *trusses* are the main structural roof supports. The *ridge* of a roof is the uppermost structural member of a roof. The *top plate* is the horizontal structural member on top of the wall studs. *Common rafters* run form the plate to the ridge. A *hip rafter* is a rafter at the intersection of two slopes of a roof that form an external angle.

Dimensions used in calculating rafter lengths are illustrated. *Line length* is the hypotenuse of a right triangle, the base of which is the run, and the height of which is the rise. *Overhang* is the horizontal distance the rafter projects beyond the wall on which it rests. The *tail* is the distance along the center line of the rafter that extends beyond the wall.

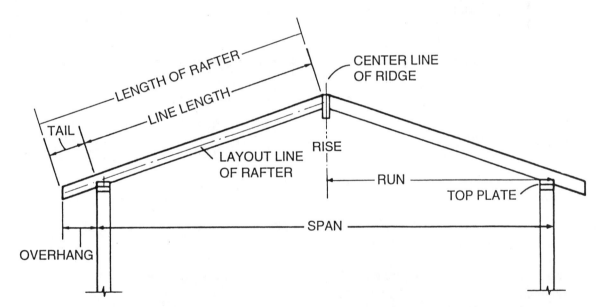

Carpenters normally use the rafter table on the steel square to determine the rafter length. In special cases, the carpenter might use the Pythagorean theorem. Rafter stock is sold in lengths from 8 feet to 16 feet in multiples of 2 feet. When estimating the number of rafters for a common gable roof, remember to add one rafter to the total for each side of the roof to allow for the first rafter on each side.

Example: What is the length of the common rafter of a gable roof that has an 8-inch rise per foot of run. The roof span is 24 feet and the overhang is 1 foot.

Solution: Rafter run = 24 feet ÷ 2 = 12 feet

Rafter run and overhang = 12 feet + 1 foot = 13 feet

Using the 8-inch rise on the steel square, the length of the common rafter is 14.42 inches per foot of run.

Length of common rafter = 14.42 inches/foot × 13 feet × 1 foot/12 inches = 15.62 feet

Note: Use the first row of numbers under the inch markings for common rafters.
Use the second row for hip rafters.

1'2	1'1	1'0	9	8	7	6	5	4	3	2	1
16 97	16 28	15 62	15	14 42	13 89	13 42	13	12 65	12 37	12 16	
20 78	20 22	19 70	19 21	18 76	18 36	18	17 69	17 44	17 23	17 09	
22 5/8	21 11 16	20 13/16	20	19 1/4	18 1/2	17 7/8	17 5/16	16 7/8	16 1/2	16 1/4	
33 15/16	32 9 16	31 1/4	30	28 7/8	27 3/4	26 13/16	26	25 5/16	24 3/4	24 5/16	
8 1/2	8 7/8	9 1/4	9 5/8	10	10 3/8	10 3/4	11 1/16	11 3/8	11 5/8	11 13/16	
9 7/8	0 1/8	10 3/8	10 5/8	10 7/8	11 1/16	11 5/16	11 1/2	11 11/16	11 13/16	11 15/16	

2'3	2'2	2'1	2'0	1'9	1'8	1'7	1'6	1'5	1'4	1'3	1'2
LENGTH	COMMON	RAFTERS	PER FOOT	RUN	21 63	20 81	20	19 21	18 44	17 69	16 97
"	HIP OR	VALLEY	" "	"	24 74	24 02	23 32	22 65	22	21 38	20 78
DIFF	IN LENGTH	OF JACKS	16 INCHES	CENTERS	28 7/8	27 3/4	26 11/16	25 5/8	24 9/16	23 9/16	22 5/8
"	"	"	2 FEET	"	43 1/4	41 5/8	40	38 7/16	36 7/8	35 3/8	33 15/16
SIDE	CUT	OF	JACKS	USE	6 11/16	6 15/16	7 3/16	7 1/2	7 13/16	8 1/8	8 1/2
"	"	HIP OR	VALLEY	"	8 1/4	8 1/2	8 3/4	9 1/16	9 3/8	9 5/8	9 7/8

PRACTICAL PROBLEMS

Note: Lumber for rafters is purchased in 2-foot multiples. Use the rafter tables above to solve problems 1–7.

1. The roof illustrated at the top of page 147 has a rise of 6 inches for every foot of run and an overhang of 1 foot.

 a. Find the length of the common rafters. a. _____

 b. Determine the number of board feet of stock required for all of the common rafters in the roof. Stock used for rafters is 2 inches by 6 inches and is spaced 16 inches o.c. b. _____

A = TAIL LENGTH
B = OVERHANG LENGTH

2. The gable roof of a house, having a span of 24 feet and length of 36 feet, has a rise of 5 inches per foot of run. The rafters are 2 × 8-inch stock, spaced 16 inches o.c., and have a 12-inch overhang.

 a. Find the length of the common rafters. a. _____

 b. Determine the total number of common rafters needed for the roof. b. _____

 c. Calculate the number of board feet of stock required for the common rafters. c. _____

3. Find the number of rafters required for a gable roof that has a 12-inch rise per foot of run, a span of 26 feet, and a roof length of 44 feet. The rafters are spaced 16 inches on center. Allow for an overhang of 1 foot. _____

Note: Use this illustration for problems 4–7.

FRAMING PLAN OF HIP ROOF

4. Find the length of the hip rafter if the roof shown has a rise of 4 inches for each foot of run. (Note that the run plus overhang for the hip rafter is 8 feet.)

5. What is the total length of hip rafters to be ordered for the roof shown if it has a 6-inch rise for each foot of run?

6. What is the length of all the hip rafters needed for the roof shown if the roof shown has a rise of 9 inches for each foot of run?

7. What is the total length of the hip rafters shown if the roof has a rise of 5 inches for each foot of run?

Unit 37 SHEATHING

BASIC PRINCIPLES OF SHEATHING

Sheathing is the structural covering over the exterior side of joists, studs, or rafters. The most commonly used materials are plywood and insulation board.

Sheet materials, such as plywood or insulation board, are sold by the square foot, commonly in 4 × 8-foot sheets. The amount required is estimated from the number of square feet to be covered.

Board sheathing is still sometimes used. An estimate of the amount required is computed by determining the number of square feet of wall area to be covered. From this amount, deduct the area of all openings (such as doors and windows) and add a percentage for waste (25% of whole area for horizontal or 30% of whole area for diagonal). The result will indicate the number of square feet of material to be ordered.

The number of roof boards required is also calculated on the basis of the square footage of roof area to be covered. If 1 × 6-inch or 1 × 8-inch boards are used, an allowance of 20% is added for waste.

The illustrations that follow indicate the usual methods of applying sheathing materials.

INSTALLATION OF BOARD SHEATHING

TOP PLATES

STUD

INSULATING SHEATHING

PLYWOOD

INSTALLATION OF PLYWOOD OR INSULATING BOARD SHEATHING

PRACTICAL PROBLEMS

1. How many square feet of 1 × 8-inch board sheathing are required to diagonally sheath a wall 8 feet by 26 feet? Make no allowance for openings. _____

2. A building measuring 8 feet by 26 feet by 42 feet is to be sheathed using horizontal 1 × 8-inch boards. Allowing a total of 187 square feet for openings, how many square feet of sheathing are required? _____

3. Plywood is used as sheathing for the exterior walls of a home that measures 28 feet by 48 feet by 8 feet. How many 4 × 8-foot sheets of ½-inch plywood are needed? _____

4. A wall 37 feet long is to be sheathed to a height of 12 feet. How many square feet of structural insulation board are required? _____

5. A gable roof is to be covered with plywood. If each side measures 20 feet by 60 feet, what is the total number of square feet of plywood that must be ordered? _____

6. How many 4 × 8-foot sheets of plywood sheathing are required for a shed roof measuring 28 feet by 16 feet? _____

Unit 38 TRIM

BASIC PRINCIPLES OF TRIM

Trim is finish material applied around the openings, ceilings, and floors of a building. Trim required for a house usually falls into two categories, exterior and interior. Most trim material is available in standard lengths and is sold by the linear foot.

Exterior trim is that portion of the exterior finish of a house other than the wall covering. Exterior trim typically includes window and door trim, as well as those items making up the cornice.

FASCIA

SOFFIT

CROWN MOLDING

FRIEZE BOARD AND MOLDING

SIDING

CORNICE INSTALLATION

Interior trim includes items such as door trim (stops and casing), window trim (casing, apron, stool), base trim (baseboard, base shoe, base cap), ceiling trim, and other items such as chair rail or picture moldings. In all cases, the estimator must find the number of linear feet of trim required, usually by determining the perimeter of the door, window, or room to be trimmed.

BASE CAP

BASEBOARD

BASE SHOE

BASE TRIM INSTALLATION

PRACTICAL PROBLEMS

1. Find the number of feet of frieze board required for a gable end cornice if the span is 24 feet and the roof pitch is ¼. Express the answer to the nearest whole foot.

FRIEZE
BOARD

2. What is the actual number of feet of baseboard required for a room measuring 12 feet by 22 feet? Deduct 8 feet for openings.

3. How many feet of window casing are required for two windows, each measuring 3 feet by 4 feet 6 inches? Allow 1 foot per window for waste. Do not allow for the stool or apron.

.OUTSIDE CASING

JAMB

SASH

PLASTER OR
DRY WALL

CASING

STOP

SILL

STOOL

APRON

TYPICAL WINDOW TRIM INSTALLATION

4. What amount of door casing is required to trim a passage door measuring 2 feet 6 inches by 6 feet 8 inches? Both sides of passage doors require casing. _____

STUDS

PLASTER OR DRY WALL

NAIL

SHINGLE WEDGE

JAMB

STOP

CASING

INSTALLATION OF DOOR JAMB AND CASING

Unit 39 ROOFING

BASIC PRINCIPLES OF ROOFING

The most common coverings for pitched roofs are asphalt and wood shingles. Other types of roofing such as tile, slate, metal, or built-up roofing are also used, but these types are not usually installed by the carpenter.

Shingles are commonly bought by the square (100 square feet). To estimate the number of squares, calculate the total number of square feet of roof surface to be covered and divide by 100 square feet. For plain roofs, add an allowance of 8% of the area for waste. A square of 18-inch shingles, laid 6 inches to the weather, covers 100 square feet. If the same shingles are laid with an exposure of 5 inches to the weather, they would only cover ⅚ of the area. Therefore, more shingles would be required to cover one square.

Asphalt shingles are also estimated on the basis of the square. Again, the total roof area is divided by 100 square feet to give the number of squares that must be ordered. A figure of 5% added for waste is often used.

INSTALLATION OF WOOD SHINGLES

PRACTICAL PROBLEMS

1. Determine the number of bundles of asphalt shingles required for both sides of a gable roof house with a span of 26 feet (including overhang), a ridge of 40 feet, and 6 inches rise per foot of run. Each bundle contains z\square. No allowance is to be made for waste. (See page 146 for rafter tables.) _____

FELT UNDERLAY

2"–4" LAP

ROOF SHEATHING
(SHOWN AS PLYWOOD)

EXPOSURE

SHEATHING

STARTING COURSE
(DOUBLE)

INSTALLATION OF ASPHALT SHINGLES

2. How many bundles of wood shingles laid 6 inches to the weather are required for one side of a gable roof measuring 16 feet by 32 feet? There are four bundles per square. Allow 8% for waste. _____

3. If asphalt shingles that weigh 235 pounds per square are used, find the weight of shingles for an area of 10 feet by 15 feet. Do not allow for waste. _____

4. a. How many squares of asphalt shingles are needed to cover two sides of a roof with a rafter length of 14 feet and a ridge 28 feet long? Allow 5% for waste.

 a. _____

 b. If asphalt shingles that weigh 235 pounds per square are used, find the total weight of the shingles used. Do not consider the weight of the 5% waste allowance.

 b. _____

5. Find the number of squares of asphalt shingles needed for two sides of a gable roof with 8 inches rise per foot of run, a span of 22 feet, and a ridge 30 feet long. (See page 146 for rafter tables.) _____

6. How many courses of shingles are required if the rafter length is 16 feet and they are laid 5 inches to the weather? _____

7. A roll of roofing paper covers an area of 100 square feet. How many rolls are needed for a shed roof measuring 12 feet by 16 feet? Do not allow for waste. _____

Unit 40 DOORS AND WINDOWS

BASIC PRINCIPLES OF DOORS AND WINDOWS

Windows, exterior doors, and their frames are items that are usually made at a factory and installed, ready for use, at the construction site. Most architectural plans contain specification sheets or a door and window schedule from which the carpenter may simply count and order the specified types.

The most common window types used in residential construction are double hung, casement, hopper, awning, fixed, and horizontal sliding. They may be made of wood or metal and usually can be ordered with regular or insulated glass and with premade screens or storm window units.

DOUBLE HUNG AWNING SLIDING

HOPPER CASEMENT

Interior doors and door frames are often purchased as completed units ready for installation. The carpenter usually frames the rough opening about 3 inches higher than the door height and about 2½ inches more than the door width in order to provide for jambs and leveling within the opening. The most common types of doors are paneled and flush. Folding doors often used for closets are flush, louvered, or paneled.

EXTERIOR DOORS

FLUSH

PANELED

PANELED

FLUSH

COMBINATION

INTERIOR DOORS

PANELED

FLUSH

LOUVERED-FOLDING

PRACTICAL PROBLEMS

Note: Use this chart for problems 1 and 2.

WINDOW SCHEDULE							
Symbol	Item	Quantity	Type Window	Rough Opening	Unit Size	Manufac- turer No.	Remarks
A	1	1	Bay	9'–9½" × 5'–4"	9'–6¼" × 4'–11⅜"	Jones 1580	Radius 10'–4½"
B	2	4	Sliding	3'–7¼" × 3'–6¼"	3'–5½" × 3'–3½"	Smith 16285	Removable Sash
C	3	1	Double Hung	2'–10⅛" × 3'–1⅛"	2'–8" × 2'–10"	Jones 1437	Insulating Glass
D	4	6	Double Hung	3'–2⅛" × 4'–5⅛"	3'–0" × 4'–2"	Jones 1467	Insulating Glass
E	5	2	Casement	5'–4" × 1'–8"	4'–11" × 1'–3"	Smith 782A	Insulating Glass

1. From the window schedule illustrated, determine the total cost of item 4 if one unit costs $124.96. _____

2. According to the window schedule illustrated, how many casement windows must be ordered? _____

Note: Use this chart for problems 3–5.

DOOR SCHEDULE						
Symbol	Quantity	Type Door	Rough Opening	Unit Size	Manufacturer No.	Remarks
A	1	Panel	3'–3½" × 6'–11½"	3'–0" × 6'–8"	Miller 821P	Exterior
B	5	Flush	2'–9⅛" × 6'–11⅛"	2'–6" × 6'–8"	Jones 1175	Passage
C	2	Panel	2'–3⅛" × 6'–11¼"	2'–0" × 6'–8"	Miller 850I	Closet
D	1	Combination	3'–3½" × 6'–11½"	3'–0" × 6'–8"	Smith 2750	Exterior
E	2	Folding	5'–3¼" × 6'–11½"	5'–0" × 6'–8"	Miller 875F	Storage
F	2	Sliding	4'–3½" × 6'–11½"	4'–0" × 6'–8"	Jones 1190	Closet

3. One unit of item B costs $98.55. Using the door schedule shown, find the total cost for all the flush-type doors. _____

4. What size rough opening is required for an interior door measuring 2 feet 6 inches by 6 feet 8 inches? _____

5. Using the door schedule, find the total cost of sliding closet doors at a unit cost of $77.51. _____

Unit 41 SIDING

BASIC PRINCIPLES OF SIDING

Siding is the exterior wall finish applied to a building. Clapboards and other types of siding are sold by the square foot. The surface area to be covered plus a percentage for lapping and waste is calculated to determine the number of square feet required for an area.

The various types of siding shown below are usually made of 1 × 6-inch or 1 × 8-inch boards. Required quantities are computed on the same basis as sheathing, although an allowance must be made for both lap and waste. For the so-called 1 × 6-inch novelty siding, the lap and waste allowance is 25%. Builder's felt is often applied to the sheathing before the siding is put on. When estimating felt, 10% is added to the area to be covered. A roll of felt usually contains 400 square feet.

TYPES OF SIDING

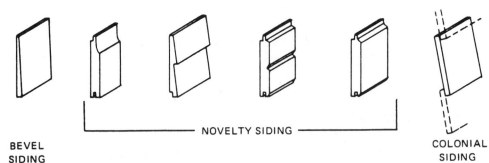

BEVEL SIDING — NOVELTY SIDING — COLONIAL SIDING

The table below gives common allowances for lap and waste on bevel siding.

TABLE OF LAP AND WASTE ALLOWANCES—BEVEL SIDING		
Width	**Exposure to Weather**	**Add for Lap and Waste**
12"	10½"	15%
12"	10"	20%
10"	8½"	18%
10"	8"	24%
8"	6¾"	22%
8"	6½"	23%
8"	6"	30%
6"	5"	25%
6"	4¾"	32%
6"	4½"	38%

PRACTICAL PROBLEMS

Note: Round answers up to next whole foot.

1. If applied as shown below, how many square feet of 6-inch bevel siding are required to cover the side of a house 24 feet by 28 feet by 18 feet 6 inches? Exposure to the weather is 4½ inches. _____

NO SHEATHING PAPER
REQUIRED FOR PLYWOOD
OR INSULATION BOARD
SHEATHING

FLASHING

BUTT JOINTS
MADE OVER STUD

STARTING STRIP

INSTALLATION OF BEVEL SIDING

2. Determine the number of square feet of 8-inch bevel siding needed to cover the sides of a house 28 feet by 32 feet by 18 feet if the exposure to the weather is 6½ inches. _____

3. For a house 26 feet by 40 feet by 17 feet 6 inches, estimate the number of square feet of 10-inch bevel siding needed. Exposure to the weather is 8½ inches. _____

4. What is the number of square feet of 8-inch bevel siding needed to cover the sides of a house 28 feet by 42 feet by 10 feet? Exposure to the weather is 6½ inches. _____

5. A garage is to be covered with bevel siding. The two sides measure 22 feet by 9 feet, and the one end, 20 feet by 9 feet. Estimate the number of square feet of 6-inch bevel siding needed if the exposure is 5 inches to the weather. _____

6. How many square feet of 8-inch bevel siding, laid 6¾ inches to the weather, are needed for a house 26 feet by 40 feet by 12 feet? _____

7. Find the number of square feet of drop siding required for two gable ends of a building with an 18-foot span and ⅓ pitch. Allow 25% for waste. (See page 95 to review pitch.) _____

8. Determine the number of rolls (400 square feet per roll) of builder's felt needed to cover the walls of a house measuring 28 feet by 46 feet by 9 feet. _____

Unit 42 STAIRS AND INTERIOR DOOR JAMBS

BASIC PRINCIPLES OF STAIRS AND INTERIOR DOOR JAMBS

Specifications that accompany each set of plans provide the necessary information on types and kinds of material to be used. The specifications also include detailed information on all standard or prefabricated parts. Examination of the plans provides any additional necessary information on special sizes.

Door jambs are ordered according to the number required; the kind of material; the width, thickness, and type of jamb; and the door size (width, height, and thickness).

Example: Method of listing door jambs:

No.	Material	Size of Jambs	Type	Door Size
6	Pine	1″ × 5 1/2″	Rabbetted	2′-6″ × 6′-8″ × 1 3/8″

INTERIOR DOOR JAMB
RABETTED

INTERIOR DOOR JAMB
WITH SEPARATE STOPS

Stair material specifications must include several important items. These are the sizes of risers, treads, stringers, rails, and newel posts. Normally the sum of the measurements of tread and riser is 17½ inches.

Example: What is the size of each tread and riser for the stairs shown?

Solution: 1. Riser height = 6' 10½" ÷ 11 risers
$$= 82.5" ÷ 11$$
$$= 7.5" \text{ or } 7½"$$

2. Tread run = 8' 4" ÷ 10
$$= 100" ÷ 10$$
$$= 10"$$

PRACTICAL PROBLEMS

1. How many risers are required for a set of stairs if the total height is 9 feet ¾ inch and the unit rise is 7¼ inches? _____

STAIR DETAIL PARTITION DETAIL

2. The unit of rise of a set of stairs is 7 inches. How many risers are needed to attain a height of 9 feet 4 inches? _____

3. Find the unit rise and unit run of a stair to fit an opening that has a total run of 9 feet 7½ inches and a rise of 7 feet. Eleven treads are used. _____

4. Find the number of treads for a flight of stairs that has a total rise of 105 inches and a unit rise of 7 inches. _____

5. How many 7⁹⁄₁₆-inch risers are there in a flight of stairs when the total rise of the stairs is 10 feet 1 inch? _____

6. How many linear feet of ¾-inch × 5½-inch lumber are required for a header and side jambs of four doors 2 feet 6 inches by 6 feet 8 inches? Allow an extra 2 inches for each side jamb. _____

7. What width door jamb is needed to frame an opening with studs measuring 1½ inches by 3½ inches, with ½-inch drywall on each side of the studs? _____

Glossary

Asbestos shingles—Fireproof shingles made from asbestos and portland cement.

Asphalt—A black mineral pitch used for waterproofing roads, driveways, walks, and to impregnate felt when used as roofing and building paper.

Asphalt shingles—Shingles made from asphalt-saturated felt and coated on the surface with various colored minerals. They are manufactured both in strips and individual shingles.

Backfill—Dirt placed in the trench around a foundation.

Baffles—Plates used to regulate the flow of materials between adjacent compartments.

Baseboard—Finishing board used at the base of a vertical wall.

Batter board—The temporary frame used to locate foundation corners.

Beam—An inclusive term for joists, girders, and rafters.

Bed molding—Molding used to cover the joint between the plancier and frieze.

Bevel siding—Wedge-shaped boards used as finish siding on the exterior of a structure.

Board foot—The equivalent of a board 1 inch thick, 1 foot wide, and 1 foot long. It equals 144 cubic inches.

Casement sash—Windows that are hinged on one side.

Casing—The trim around a door or window, either outside or inside. The finish lumber covering a rough-sawed post is also called casing.

Cement—A fine grayish powder produced by burning in a kiln a carefully blended mixture of limestone with clay, or shale. The clinkers thus formed are pulverized and bagged.

Chair rail—Horizontal moldings placed on a wall to protect the wall from damage by furniture.

Circumference—The perimeter of a circle.

Cistern—A large tank for holding water.

Clapboards—Boards used as exterior siding, placed horizontally and lapped.

Clothes chute—A chute designed for the purpose of dropping soiled clothes from an upper level to a basement laundry.

Collar beam—A beam above the wall plate used to tie two rafters together.

Common rafter—Main structural roof support running from the wall plate to the ridge at right angles to both plate and ridge.

Concrete—A building material made from a mixture of cement, sand, gravel, and water.

Corner mold—Molding applied to exterior corners of a room to protect the plaster or drywall board.

Cornice—The section of exterior trim from the top of the wall to the projection of the rafters.

Crown molding—A molding used to conceal the joint formed by the ceiling and wall. It is placed in contact with both ceiling and wall.

Culvert—Arched channel for the passage of water below ground level.

Dado—A rectangular cut in a piece of wood across the wood grain.

Dead load—The weight of all stationary construction included in a building.

Diameter—A line passing through the center of a circle and terminating at both ends by the circumference. It is equal in distance to twice the radius.

Drywall—A wall covering of stone or other durable material, laid without mortar. Gypsum board is a common type of drywall.

Fascia—A board nailed to the ends of the rafters.

Fiberboard—Sheet material formed from cellulose or wood fibers pressed together.

Fire stop—A block or closure between the studs or joists to prevent the spread of fire or smoke; also called draft stop.

Flashing—Material used around chimneys, vents, windows, and so on to prevent moisture from entering the building. Sheet metal is a very common type of flashing.

Flush—Two members forming an even surface.

Footing—Enlargement at the base of a pier, column, or foundation wall.

Form—A wood or metal structure constructed for the purpose of retaining fresh concrete until it sets.

Foundation—The supporting portion of a structure usually below grade. It is composed of two parts, footing and foundation wall.

Framing square—A right angle tool used in framing buildings. It is generally made of steel and includes tables in addition to the scales.

Framing timber—Lumber used in framing with dimensions greater than 4 inches by 6 inches.

Frieze—Decorative trim on the upper part of a wall connecting the top of the siding with the soffit.

Gable—The portion of a wall bound by the plate and the inclined roof of a building.

Gable roof—A roof formed by common rafters. It slopes up from only two walls.

Gambrel roof—A gable roof each slope of which is broken into two planes with different pitches.

Girder—A large horizontal beam used to support smaller beams or joists.

Glazing—The process of installing glass into a sash frame.

Glazing point—Small piece of metal used to secure glass in a window prior to puttying.

Grout—A watery mixture of sand and cement.

Hardboard—Pressed wood panels made from wood fibers, not wood chips. It is normally bonded by the natural materials in wood.

Header—In floor framing: A short traverse or horizontal joist that supports the ends of one or more cut-off joists. In wall framing: Supporting member for the load over the opening for a window or door; also called lintel.

Hip rafter—A rafter at the intersection of two roof slopes forming an external angle or a hip. It extends diagonally from the corner of the plate to the ridge.

Hip roof—A roof that slopes up toward the center or ridge from four sides.

Horizontal—Left-and-right; parallel to the horizon.

Hypotenuse—The longest side of a right triangle. It is the side opposite the 90° angle.

Isosceles triangle—A triangle with two sides of equal measure.

Jamb—The vertical posts of door or window framing; also frequently used to refer to the vertical casing within a doorway or window.

Joint—The place where two or more members come together.

Joist—Beam used to support a floor or ceiling.

Laminated plastic—Layered plastic bound with resin, frequently used as a surface on countertops.

Lath—Wood strip, metal mesh, or gypsum used as a foundation for plaster.

Legs of triangle—The sides of a right triangle that form the right angle.

Linear foot—A measure of twelve inches along a straight line.

Line length—The hypotenuse of a triangle whose base is the total run and whose altitude is the total rise of a roof.

Live load—Movable load, such as people, furnishings, and so on, as contrasted to dead load, which is a permanent part of a building.

Load-carrying partitions—Partitions that support the floor joists above them.

Louver—Ventilation opening.

Masonry—Materials bonded together with mortar to form a wall.

Masonry cement—A mixture of Portland cement and lime.

Matched boards—Boards that are edge dressed and shaped into tongue-and-groove edges to join together snugly.

Miter box—A device used to guide a saw when cutting at an angle.

Mortise—A hole or slot into which some other part or tenon fits.

Newel post—A vertical post at the foot of a stairway into which the handrail is fitted.

Novelty siding—Siding that has been processed to give an unusual design. This siding can serve as both siding and sheathing.

On center (o.c.)—A term that refers to the distance from the center of one structural member to the center of the next member.

Overhang—The lower part of a roof that projects over the exterior wall. It is also the horizontal distance that a rafter extends beyond a wall.

Particle board—Wood chips with a resin binder pressed into sheets of various thicknesses.

Penny—A term used to indicate the length of nail. The letter "d" is used as an abbreviation for the term penny.

Perimeter—The length of the outer boundary of a plane figure.

Pier—A piece of solid upright masonry designed to support a structural load.

Pitch (roof)—The incline of a roof expressed as rise divided by the span.

Plate—A horizontal structural member to which the top and bottom of the studs are secured.

Plywood—Boards made of several layers of wood. The grain of alternate pieces is at right angles.

Potter's wheel—A horizontal wheel activated by the potter's feet, thus rotating the clay.

Projection—The line length of rafter that extends beyond the plate.

Radius—Distance from the center of a circle to any point on the circumference.

Rafter—Main structural roof support.

Railing (stair)—A complete assembly consisting of handrail, balustrade, and newel post.

Retaining wall—Concrete or masonry wall built to support the earth in a sloping area.

Ribbon—A narrow board secured to the edge of studs to support the joists.

Ridge—The uppermost horizontal portion of a roof.

Right triangle—A triangle with a 90° angle.

Rise—The change in elevation of a roof or set of stairs.

Roof boards—Boards that cover the exterior of the roof rafters.

Run—The distance from the face of the upper stair riser to the face of the lower stair riser. It is also the horizontal distance covered by a rafter.

Saw kerf—The width of a cut or incision made by a saw in a piece of wood.

Scaffold—A temporary platform used to support workers and equipment.

Semicircular—Shape of one-half of a circle.

Septic tank—An underground metal, masonry, or wooden tank for retaining solid sewage that is then decomposed by bacteria.

Sheathing—Lumber or man-made sheets placed on the exterior of the studs or rafters.

Shim—A thin, often tapered, strip of wood or metal used to fill in a space when leveling.

Shingle—Wood, asphalt, or asbestos units that are applied to the roof or sidewalls to make the surface weathertight.

Shoe molding—The molding that conceals the joint between the floor and baseboard.

Siding—The exterior covering of the sides of a building.

Sill—The lowest horizontal member of a frame structure, usually resting upon a foundation. It is also the lowest member of a door or window frame.

Slope—The incline of a roof expressed as a ratio of rise to run.

Soffit—The underside of the members of a building, such as the undersurface of the cornice.

Sole plate—The horizontal member on the subfloor to which studs are nailed.

Span—The distance between opposite sides of a building.

Specifications—A written statement or part of the contract listing the kind and quality of materials or workmanship used in the construction.

Square—A quantity of material sufficient to cover 100 square feet of surface.

Square foot—A unit of area equal to 144 square inches.

Stair riser—The vertical boards located between treads of a set of stairs.

Stair tread—The horizontal boards upon which one steps.

Story pole—A stick or rod used to locate elements in construction. Stair risers, courses of siding, height or location of framing members, and cabinet layout.

Stringer—The inclined member on each side of a set of stairs to which the treads and risers are secured, usually made of finished wood.

Stud—The vertical members of outside walls or partitions.

Subfloor—Plywood or matched boards laid directly on the floor joists upon which a finished floor is laid.

Tail—The portion of a rafter that extends beyond a wall.

Tenon—A projecting member left by cutting away the wood around it for insertion into a mortise to make a joint.

Tongue and groove (T and G)—Boards that are joined by cutting a tongue (rib) on the edge of one board and a groove on the edge of an adjacent board.

To the weather—A term that refers to the exposed portion of siding or shingle.

Trapezoid—A quadrilateral with one and only one pair of sides parallel.

Trim—The finishing materials in a building such as molding, baseboard, cornice, and so on.

Vertical—Up-and-down; upright, perpendicular to a horizontal.

Wallboard—Gypsum, wood pulp, or similar materials made into large sheets that are nailed to the interior of a building frame to form interior walls.

Window apron—Member of the trim of a window located on the wall immediately beneath the stool.

Window stool—A horizontal member placed on top of a window sill on the interior of a window.

Appendix

SECTION I

DENOMINATE NUMBERS

Denominate numbers are numbers that include units of measurement. The units of measurement are arranged from the largest unit at the left to the smallest unit at the right.

Example: 6 yd 2 ft 4 in

All basic operations of arithmetic can be performed on denominate numbers.

I. EQUIVALENT MEASURES

Measurements that are equal can be expressed in different terms. For example, 12 in = 1 ft. To express one measurement as another equal measurement, multiply by the equivalent measurement in the form of 1. To express 6 inches in equivalent foot measurement, multiply 6 inches by 1 in the form of the conversion factor

$$\frac{1\,ft}{12\,in}$$

In the numerator and denominator, divide by a common factor.

$$6\,in = \frac{\overset{1}{\cancel{6\,in}}}{1} \times \frac{1\,ft}{\underset{2}{\cancel{12\,in}}} = \frac{1}{2}\,ft \text{ or } 0.5\,ft$$

To express 4 feet in equivalent inch measurement, multiply 4 feet by 1 in the form of the conversion factor

$$\frac{12\,in}{1\,ft}$$

$$4\,ft = \frac{\overset{4}{\cancel{4\,ft}}}{} \times \frac{12\,in}{\underset{1}{\cancel{1\,ft}}} = \frac{48\,in}{1} = 48\,in$$

Per means division, as with a fraction bar. For example, 50 miles per hour can be written

$$\frac{50 \text{ miles}}{1 \text{ hour}}$$

II. BASIC OPERATIONS

A. Addition

Example: 2 yd 1 ft 5 in + 1 ft 8 in + 5 yd 2 ft

1. Write the denominate numbers in a column with like units in the same column.

	2 yd	1 ft	5 in
		1 ft	8 in
+	5 yd	2 ft	

2. Add the denominate numbers in each column.

	7 yd	4 ft	13 in

3. Express the answer using the largest possible units.

7 yd			=	7 yd		
	4 ft		=	1 yd	1 ft	
		13 in	= +		1 ft	1 in
7 yd	4 ft	13 in	=	8 yd	2 ft	1 in

B. Subtraction

Example: 4 yd 3 ft 5 in − 2 yd 1 ft 7 in

1. Write the denominate numbers in columns with like units in the same column.

	4 yd	3 ft	5 in
−	2 yd	1 ft	7 in

2. Starting at the right, examine each column to compare the numbers. If the bottom number is larger, exchange one unit from the column at the left for its equivalent. Combine like units.

7 in is larger than 5 in, so borrow 12 in from 3 ft and add it to 5 in

3 ft = 2 ft 12 in

12 in + 5 in = 17 in

3. Subtract the denominate numbers.

	4 yd	2 ft	17 in
−	2 yd	1 ft	7 in
	2 yd	1 ft	10 in

4. Express the answer using the largest possible units.

	2 yd	1 ft	10 in

C. Multiplication

—By a constant

Example: 1 hr 24 min × 3

1. Multiply the denominate number
 by the constant.

2. Express the answer using the
 largest possible units.

$$\begin{array}{ll} \text{1 hr} & \text{24 min} \\ & \underline{\times\ 3} \\ \text{3 hr} & \text{72 min} \\ \downarrow & \quad \downarrow \\ \text{3 hr} & \qquad\qquad = \text{3 hr} \\ & \text{72 min} = \text{1 hr} \quad \text{12 min} \\ \hline \text{3 hr} & \text{72 min} = \text{4 hr} \quad \text{12 min} \end{array}$$

—By a denominate number expressing linear measurement

Example: 9 ft 6 in × 10 ft

1. Express all denominate numbers
 in the same unit.

 $$9 \text{ ft } 6 \text{ in} = 9\frac{1}{2} \text{ ft}$$

2. Multiply the denominate numbers.
 (This includes the units of measure,
 such as ft × ft = sq ft.)

 $$9\frac{1}{2} \text{ ft} \times 10 \text{ ft} =$$

 $$\frac{19}{2} \text{ ft} \times 10 \text{ ft} = 95 \text{ sq ft}$$

—By a denominate number expressing square measurement

Example: 3 ft × 6 sq ft

1. Multiply the denominate numbers.
 (This includes the units of measure,
 such as ft × ft = sq ft and sq ft × ft = cu ft.)

 3 ft × 6 sq ft = 18 cu ft

—By a denominate number expressing rate

Example: 50 mi per hr × 3 hr

1. Express the rate as a fraction using
 the fraction bar for *per*.

 $$\frac{50 \text{ mi}}{1 \text{ hr}} \times \frac{3 \text{ hr}}{1}$$

2. Cancel any common factors. (This may include units of measure, such as hr ÷ hr = 1.)

$$\frac{50 \text{ mi}}{\overset{}{\underset{1}{\cancel{1\text{ hr}}}}} \times \frac{\overset{3}{\cancel{3\text{ hr}}}}{1}$$

3. Multiply numerators, expressing the answer in the remaining unit.

$$\frac{150 \text{ mi}}{1} = 150 \text{ mi}$$

D. Division

—By a constant

Example: 8 gal 3 qt ÷ 5

1. Express all denominate numbers in the smallest unit.

8 gal 3 qt = 35 qt

2. Divide the denominate number by the constant.

35 qt ÷ 5 = 7 qt

3. Express the answer using the largest possible units.

7 qt = 1 gal 3 qt

—By a linear measurement

Example: 11 ft 4 in ÷ 8 in

1. Express all denominate numbers in the same unit.

11 ft 4 in = 136 in

2. Divide the denominate numbers by cancelling common factors. (This includes the units of measure, such as in ÷ in = 1.)

136 in ÷ 8 in =

$$\frac{\overset{17}{\cancel{136 \text{ in}}}}{\underset{1}{\cancel{8 \text{ in}}}} = \frac{17}{1} = 17$$

—By a linear measure with a square measurement as the dividend

Example: 20 sq ft ÷ 4 ft

1. Divide the denominate numbers.
 (This includes the units of measure,
 such as sq ft ÷ ft = ft.)

2. Express the answer in the
 remaining unit.

$$20 \text{ sq ft} \div 4 \text{ ft}$$

$$\frac{\overset{5 \text{ ft}}{\cancel{20 \text{ sq ft}}}}{\underset{1}{\cancel{4 \text{ ft}}}} = \frac{5 \text{ ft}}{1} = 5 \text{ ft}$$

—By denominate numbers used to find rate

Example: 200 mi ÷ 10 gal

1. Divide the denominate numbers.

$$\frac{\overset{20 \text{ mi}}{\cancel{200 \text{ mi}}}}{\underset{1 \text{ gal}}{\cancel{10 \text{ gal}}}} = \frac{20 \text{ mi}}{1 \text{ gal}}$$

2. Express the units with the fraction
 bar meaning *per*.

$$\frac{20 \text{ mi}}{1 \text{ gal}} = 20 \text{ miles per gallon}$$

Note: Alternate methods of performing the basic operations will produce the same results. The choice
 of method is determined by the individual.

III. FORMULAS

Area = Length × Width

Part = Percent × Whole or $P = \% \times W$

$\% = P / W$

$W = P / \%$

Interest = Principal × Rate × Time (in yrs)

$I = PRT$

Sale price or Net price = List price − Discount $(N = L − D)$

Double Discounts: $N = L \times D_1 \times D_2$

Circumference = Diameter × 3.1416 = 3.1416d

Diameter = Radius × 2 = 2r

Perimeter of square = 4s

Perimeter of rectangle = 2l + 2w

Area of square = s^2

Area of rectangle = $l \times w$

Area of triangle = ½ × *base × height* = ½ × *bh*

Area of circle = 3.1416 × r × r = 3.1416r^2

Area of trapezoid = $(b_1 + b_2)$ ½ × h

Volume of cube = $s \times s \times s = s^3$

Volume of cylinder = 3.1416 × h × r × r = 3.1416hr^2

Volume of rectangular solid = *lwh*

Board foot = $T \times W \times L$

Run = Span/2

Span = Run × 2

Pitch = Rise/Span

Stairs: Tread + Riser (Generally have measures totaling about 17½ inches.)

SECTION II

TABLE I

STANDARD TABLES OF CUSTOMARY UNITS OF MEASURE

Linear Measure		
12 inches (in.)	=	1 foot (ft.)
3 ft.	=	1 yard (yd.)
16 1/2 ft.	=	1 rod (rd.)
5 1/2 yd.	=	1 rd.
320 rd.	=	1 mile
1760 yd.	=	1 mile
5280 ft.	=	1 mile

Surface Measure		
144 sq. in.	=	1 sq. ft.
9 sq. ft.	=	1 sq. yd.
30 1/4 sq. yd.	=	1 sq. rd.
160 sq. rd.	=	1 acre
640 acres	=	1 sq. mile
43,560 sq. ft.	=	1 acre

Cubic Measure		
1728 cu. in.	=	1 cu. ft.
27 cu. ft.	=	1 cu. yd.
128 cu. ft.	=	1 cord

Angular (Circular) Measure		
60 sec. (")	=	1 min. (')
60'	=	1 degree (°)
90°	=	1 quadrant
360°	=	1 circle

Time Measure		
60 seconds (sec.)	=	1 minute (min.)
60 min.	=	1 hour (hr.)
24 hr.	=	1 day
7 days	=	1 week
52 weeks	=	1 year
365 days	=	1 year
10 years	=	1 decade

Liquid Measure		
4 gills	=	1 pint (pt.)
2 pt.	=	1 quart (qt.)
4 qt.	=	1 gallon (gal.)
231 cu. in.	=	1 gal.
31.5 gal.	=	1 barrel (bbl.)
42 gal.	=	1 bbl. of oil
8 1/2 lb.	=	1 gal. water
7 1/2 gal.	=	1 cu. ft.

Weights of Materials		
0.096 lb.	=	1 cu. in. aluminum
0.260 lb.	=	1 cu. in. cast iron
0.283 lb.	=	I cu. in. mild steel
0.321 lb.	=	1 cu. in. copper
0.41 lb.	=	1 cu. in. lead
112 lb.	=	1 cu. ft. Dowmetal
167 lb.	=	1 cu. ft. aluminum
464 lb.	=	1 cu. ft. cast iron
490 lb.	=	1 cu. ft. mild steel
555.6 lb.	=	1 cu. ft. copper
710 lb.	=	1 cu. ft. lead

Avoirdupois Weight		
16 ounces (oz.)	=	1 pound (lb.)
100 1b.	=	1 hundredweight (cwt.)
20 cwt.	=	1 ton
2000 lb.	=	1 ton
8 1/2 lb.	=	1 gal. of water
62.4 lb.	=	1 cu. ft. of water
112 lb.	=	1 long cwt.
2240 lb.	=	1 long ton

Dry Measure		
2 cups	=	1 pt.
2 pt.	=	1 qt.
4 qt.	=	1 gal.
8 qt.	=	1 peck (pk.)
4 pk.	=	1 bushel (bu.)

Miscellaneous		
12 units	=	1 dozen (doz.)
12 doz.	=	1 gross
144 units	=	1 gross
24 sheets	=	1 quire
20 quires	=	1 ream
20 units	=	1 score
6 ft.	=	1 fathom

Metric Measure		
Millimeter (mm) 1 mm	=	.001M
Centimeter (cm) 1 cm	=	.01M
Decimeter (dm) 1 dm	=	.1M
Meter (M) 1M	=	1M
Dekameter (dkm) 1 dkm	=	10M
Hectometer (hm) 1 hm	=	100M
Kilometer (km) 1 km	=	1000M

TABLE II

CONVERSION OF ENGLISH AND METRIC UNITS OF MEASURE

Linear Measure

Unit	Inches to millimeters	Millimeters to inches	Feet to meters	Meters to feet	Yards to meters	Meters to yards	Miles to kilometers	Kilometers to miles
1	25.40	0.03937	0.3048	3.281	0.9144	1.094	1.609	0.6214
2	50.80	0.07874	0.6096	6.562	1.829	2.187	3.219	1.243
3	76.20	0.1181	0.9144	9.842	2.743	3.281	4.828	1.864
4	101.60	0.1575	1.219	13.12	3.658	4.374	6.437	2.485
5	127.00	0.1968	1.524	16.40	4.572	5.468	8.047	3.107
6	152.40	0.232	1.829	19.68	5.486	6.562	9.656	3.728
7	177.80	0.756	2.134	22.97	6.401	7.655	11.27	4.350
8	203.20	0.3150	2.438	26.25	7.315	8.749	12.87	4.971
9	228.60	0.3543	2.743	29.53	8.230	9.842	14.48	5.592

Example 1 in. = 25.40 mm 1 m = 3.281 ft. 1 km = 0.6214 mi.

Surface Measure

Unit	Square inches to square centimeters	Square centimeters to square inches	Square feet to square meters	Square meters to square feet	Square yards to square meters	Square meters to square yards	Acres to hectares	Hectares to acres	Square miles to square kilometers	Square kilometers to square miles
1	6.452	0.1550	0.0929	10.76	0.8361	1.196	0.4047	2.471	2.59	0.3861
2	12.90	0.31	0.1859	21.53	1.672	2.392	0.8094	4.942	5.18	0.7722
3	19.356	0.465	0.2787	32.29	2.508	3.588	1.214	7.413	7.77	1.158
4	25.81	0.62	0.3716	43.06	3.345	4.784	1.619	9.884	10.36	1.544
5	32.26	0.775	0.4645	53.82	4.181	5.98	2.023	12.355	12.95	1.931
6	38.71	0.93	0.5574	64.58	5.017	7.176	2.428	14.826	15.54	2.317
7	45.16	1.085	0.6503	75.35	5.853	8.372	2.833	17.297	18.13	2.703
8	51.61	1.24	0.7432	86.11	6.689	9.568	3.237	19.768	20.72	3.089
9	58.08	1.395	0.8361	96.87	7.525	10.764	3.642	22.239	23.31	3.475

Example 1 sq. in. = 6.452 cm^2, 1 m^2 = 1.196 sq. yd., 1 sq. mi. = 2.59 km^2

Cubic Measure

Unit	Cubic inches to cubic centimeters	Cubic centimeters to inches	Cubic feet to cubic meters	Cubic meters to cubic feet	Cubic yards to cubic meters	Cubic meters to cubic yards	Gallons to cubic feet	Cubic feet to gallons
1	16.39	0.06102	0.02832	35.31	0.7646	1.308	0.1337	7.481
2	32.77	0.1220	0.05663	70.63	1.529	2.616	0.2674	14.96
3	49.16	0.1831	0.08495	105.9	2.294	3.924	0.4010	22.44
4	65.55	0.2441	0.1133	141.3	3.058	5.232	0.5347	29.92
5	81.94	0.3051	0.1416	176.6	3.823	6.540	0.6684	37.40
6	98.32	0.3661	0.1699	211.9	4.587	7.848	0.8021	44.88
7	114.7	0.4272	0.1982	247.2	5.352	9.156	0.9358	52.36
8	131.1	0.4882	0.2265	282.5	6.116	10.46	1.069	59.84
9	147.5	0.5492	0.2549	371.8	6.881	11.77	1.203	67.32

Example 1 cm^3 = 0.06102 cu. in., 1 gal. = 0.1337 cu. ft.

Volume or Capacity Measure

Unit	Liquid ounces to cubic centimeters	Cubic centimeters to liquid ounces	Pints to liters	Liters to pints	Quarts to liters	Liters to quarts	Gallons to liters	Liters to gallons	Bushels to hectoliters	Hectoliters to bushels
1	29.57	0.03381	0.4732	2.113	0.9463	1.057	3.785	0.2642	0.3524	2.838
2	59.15	0.06763	0.9463	4.227	1.893	2.113	7.571	0.5284	0.7048	5.676
3	88.72	0.1014	1.420	6.340	2.839	3.785	11.36	0.7925	1.057	8.513
4	118.3	0.1353	1.893	8.454	3.170	4.227	15.14	1.057	1.410	11.35
5	147.9	0.1691	2.366	10.57	4.732	5.284	18.93	1.321	1.762	14.19
6	177.4	0.2029	2.839	12.68	5.678	6.340	22.71	1.585	2.114	17.03
7	207.0	0.2367	3.312	14.79	6.624	7.397	26.50	1.849	2.467	19.86
8	236.6	0.2705	3.785	16.91	7.571	8.454	30.28	2.113	2.819	22.70
9	266.2	0.3043	4.259	19.02	8.517	9.510	34.07	2.378	3.171	25.54

Example 1 L = 2.113 pt., 1 gal. = 3.785 L

TABLE III

METRIC AND CUSTOMARY DECIMAL EQUIVALENTS
FOR FRACTIONAL PARTS OF AN INCH

Fraction	Decimal Equivalent		Fraction	Decimal Equivalent	
	Customary (in.)	Metric (mm)		Customary (in.)	Metric (mm)
1/64— 0.015625	0.3969		33/64—0.515625	13.0969	
1/32———— 0.03125	0.7938		17/32————0.53125	13.4938	
3/64 — 0.046875	1.1906		35/64—0.546875	13.8906	
1/16————0.0625	1.5875		9/16————0.5625	14.2875	
5/64— 0.078125	1.9844		37/64—0.578125	14.6844	
3/32————0.09375	2.3813		19/32————0.59375	15.0813	
7/64—0.109375	2.7781		39/64—0.609375	15.4781	
1/8————————0.1250	3.1750		5/8————————0.6250	15.8750	
9/64 — 0.140625	3.5719		41/64—0.640625	16.2719	
5/32————0.15625	3.9688		21/32————0.65625	16.6688	
11/64—0.171875	4.3656		43/64—0.671875	17.0656	
3/16————0.1875	4.7625		11/16————0.6875	17.4625	
13/64—0.203125	5.1594		45/64—0.703125	17.8594	
7/32————0.21875	5.5563		23/32————0.71875	18.2563	
15/64—0.234375	5.9531		47/64—0.734375	18.6531	
1/4————————0.250	6.3500		3/4————————0.750	19.0500	
17/64—0.265625	6.7469		49/64—0.765625	19.4469	
9/32————0.28125	7.1438		25/32————0.78125	19.8438	
19/64—0.296875	7.5406		51/64—0.796875	20.2406	
5/16————0.3125	7.9375		13/16————0.8125	20.6375	
21/64—0.328125	8.3384		53/64—0.828125	21.0344	
11/32————0.34375	8.7313		27/32————0.84375	21.4313	
23/64—0.359375	9.1281		55/64—0.859375	21.8281	
3/8————————0.3750	9.5250		7/8————————0.8750	22.2250	
25/64—0.390625	9.9219		57/64—0.890625	22.6219	
13/32————0.40625	10.3188		29/32————0.90625	23.0188	
27/64—0.421875	10.7156		59/64—0.921875	23.4156	
7/16————0.4375	11.1125		15/16————0.9375	23.8125	
29/64—0.453125	11.5094		61/64—0.953125	24.2094	
15/32————0.46875	11.9063		31/32————0.96875	24.6063	
31/64 – 0.484375	12.3031		63/64—0.984375	25.0031	
1/2————————0.500	12.7000		1————————1.000	25.4000	

TABLE IV

POWERS AND ROOTS OF NUMBERS (1 through 100)

Num-ber	Powers		Roots		Num-ber	Powers		Roots	
	Square	Cube	Square	Cube		Square	Cube	Square	Cube
1	1	1	1.000	1.000	51	2,601	132,651	7.141	3.708
2	4	8	1.414	1.260	52	2,704	140,608	7.211	3.733
3	9	27	1.732	1.442	53	2,809	148,877	7.280	3.756
4	16	64	2.000	1.587	54	2,916	157,464	7.348	3.780
5	25	125	2.236	1.710	55	3,025	166,375	7.416	3.803
6	36	216	2.449	1.817	56	3,136	175,616	7.483	3.826
7	49	343	2.646	1.913	57	3,249	185,193	7.550	3.849
8	64	512	2.828	2.000	58	3,364	195,112	7.616	3.871
9	81	729	3.000	2.080	59	3,481	205,379	7.681	3.893
10	100	1,000	3.162	2.154	60	3,600	216,000	7.746	3.915
11	121	1,331	3.317	2.224	61	3,721	226,981	7.810	3.936
12	144	1,728	3.464	2.289	62	3,844	238,328	7.874	3.958
13	169	2,197	3.606	2.351	63	3,969	250,047	7.937	3.979
14	196	2,744	3.742	2.410	64	4,096	262,144	8.000	4.000
15	225	3,375	3.873	2.466	65	4,225	274,625	8.062	4.021
16	256	4,096	4.000	2.520	66	4,356	287,496	8.124	4.041
17	289	4,913	4.123	2.571	67	4,489	300,763	8.185	4.062
18	324	5,832	4.243	2.621	68	4,624	314,432	8.246	4.082
19	361	6,859	4.359	2.668	69	4,761	328,509	8.307	4.102
20	400	8,000	4.472	2.714	70	4,900	343,000	8.367	4.121
21	441	9,261	4.583	2.759	71	5,041	357,911	8.426	4.141
22	484	10,648	4.690	2.802	72	5,184	373,248	8.485	4.160
23	529	12,167	4.796	2.844	73	5,329	389,017	8.544	4.179
24	576	13,824	4.899	2.884	74	5,476	405,224	8.602	4.198
25	625	15,625	5.000	2.924	75	5,625	421,875	8.660	4.217
26	676	17,576	5.099	2.962	76	5,776	438,976	8.718	4.236
27	729	19,683	5.196	3.000	77	5,929	456,533	8.775	4.254
28	784	21,952	5.292	3.037	78	6,084	474,552	8.832	4.273
29	841	24,389	5.385	3.072	79	6,241	493,039	8.888	4.291
30	900	27,000	5.477	3.107	80	6,400	512,000	8.944	4.309
31	961	29,791	5.568	3.141	81	6,561	531,441	9.000	4.327
32	1,024	32,798	5.657	3.175	82	6,724	551,368	9.055	4.344
33	1,089	35,937	5.745	3.208	83	6,889	571,787	9.110	4.362
34	1,156	39,304	5.831	3.240	84	7,056	592,704	9.165	4.380
35	1,225	42,875	5.916	3.271	85	7,225	614,125	9.220	4.397
36	1,296	46,656	6.000	3.302	86	7,396	636,056	9.274	4.414
37	1,369	50,653	6.083	3.332	87	7,569	658,503	9.327	4.481
38	1,444	54,872	6.164	3.362	88	7,744	681,472	9.381	4.448
39	1,521	59,319	6.245	3.391	89	7,921	704,969	9.434	4.465
40	1,600	64,000	6.325	3.420	90	8,100	729,000	9.487	4.481
41	1,681	68,921	6.403	3.448	91	8,281	753,571	9.539	4.498
42	1,764	74,088	6.481	3.476	92	8,464	778,688	9.592	4.514
43	1,849	79,507	6.557	3.503	93	8,649	804,357	9.644	4.531
44	1,936	85,184	6.633	3.530	94	8,836	830,584	9.695	4.547
45	2,025	91,125	6.708	3.557	95	9,025	857,375	9.747	4.563
46	2,116	97,336	6.782	3.583	96	9,216	884,736	9.798	4.579
47	2,209	103,823	6.856	3.609	97	9,409	912,673	9.849	4.595
48	2,304	110,592	6.928	3.634	98	9,604	941,192	9.900	4.610
49	2,401	117,649	7.000	3.659	99	9,801	970,299	9.950	4.626
50	2,500	125,000	7.071	3.684	100	10,000	1,000,000	10.000	4.642

ANSWERS TO ODD-NUMBERED PRACTICAL PROBLEMS

SECTION 1 WHOLE NUMBERS

UNIT 1 ADDITION OF WHOLE NUMBERS

1. 112 feet
3. 7,088
5. $3,749
7. 4,186 square inches
9. 57,474 board feet
11. 33 feet
13. 5,510 square feet
15. $7,312
17. 8,085 board feet
19. 510 square feet
21. $771
23. 5,498 board feet
25. 1,908 board feet
27. a. 150 feet
 b. 448 feet
 c. 276 feet
 d. 330 feet
 e. 552 feet
27. f. 264 feet
 g. 226 feet
 h. 190 feet
 i. 532 feet
 j. 460 feet
 k. 300 feet
 l. 292 feet
29. 174 feet

UNIT 2 SUBTRACTION OF WHOLE NUMBERS

1. 77 inches
3. 269 ounces
5. 2,486 square feet
7. 5,520 gallons
9. 13,053
11. 20,900 board feet
13. 14 inches
15. 2,766 feet
17. 475 cubic meters
19. $715
21. $855.00
23. 5 feet
25. 15 feet
27. 9 feet
29. 14 feet
31. 871 sheets

UNIT 3 MULTIPLICATION OF WHOLE NUMBERS

1. 704 inches
3. 4,064
5. 112,158 inches
7. 139,708 yards
9. $68,685
11. 1,077,249 gallons
13. 175 feet
15. 1,600 feet
17. $1,036
19. 2,340 square feet
21. 64,500 shingles
23. 2,279 board feet
25. 1,080 feet
27. a. A = 22 feet
 B = 51 feet
27. b. Section I = 528
 square feet
 Section II = 1,725
 square feet
 c. Total = 2,253
 square feet

UNIT 4 DIVISION OF WHOLE NUMBERS

1. 141 inches
3. 7,776 R 1
5. 101 centimeters
7. 28 R 92
9. 1,054 millimeters

11. 11 hours
13. 21 hours
15. 49 studs
17. 7 inches
19. 37 trusses

21. 37 joists
23. 46 joists
25. 117 joists
27. 90 piers

UNIT 5 COMBINED OPERATIONS WITH WHOLE NUMBERS

1. 2,051
3. 424
5. 3,348,660
7. 24

9. 13 inches
11. 143 hours
13. 22 pieces
15. 635 square feet

17. 16½ feet
19. 72 square feet
21. $30,250

SECTION 2 COMMON FRACTIONS

UNIT 6 ADDITION OF COMMON FRACTIONS

1. 1⅜ inches
3. 1¼ gallons
5. 2⁷/₂₄ square feet
7. 7¼ hours
9. ¹³/₁₆ inch
11. 1¾ inches

13. 4⅝ inches
15. 4½ inches
17. 3⅜ inches
19. 1⅜ inches
21. 30¾ inches (2' 7½")
23. 31½ inches (2' 7½")

25. 10³/₁₆ inches
27. 38⁵/₁₆ inches
29. ⅝ inch
31. 41¼ inches

UNIT 7 SUBTRACTION OF COMMON FRACTIONS

1. ⁹/₃₂ inch
3. ¹¹/₁₆ inch
5. ¾ inch
7. ¼ inch

9. 1 inch
11. 28¾ inches
13. 4½ inches
15. 36½ inches

17. ⅝ foot
19. 41¼ inches
21. 10¼ inches
23. ⅜ inch

UNIT 8 MULTIPLICATION OF COMMON FRACTIONS

1. ¹⁵/₆₄ inch
3. 10³/₃₂ gallons
5. 37⁸/₉
7. ³/₃₂ board foot

9. ⁷/₃₂
11. 128¼ inches
13. 11' 1"
15. 19⅛' = 19' 1½"

17. 75 inches = 6' 3"
19. 1¼ inches
21. 4½ inches
23. 11⅞ inches

25. 19 studs
27. 101¼" = 8' 5¼"
29. 709 square feet
31. 43¼ inches = 3' 7¼"

UNIT 9 DIVISION OF COMMON FRACTIONS

1. ¼
3. 4 yards
5. ⅝ gallon
7. 2⅓
9. 1¹⁰⁄₃₃ inches

11. 16 pieces
13. 8 feet
15. 12
17. 48 boards
19. 6 boards

21. 40½ feet
23. 19 studs
25. 8 risers
27. 15⁵⁄₁₆
29. 15 treads

UNIT 10 COMBINED OPERATIONS WITH COMMON FRACTIONS

1. 375 feet
3. 430.666 square feet

5. $759.00

7. 89 inches or 7' 5"

SECTION 3 DECIMAL FRACTIONS

UNIT 11 ADDITION OF DECIMAL FRACTIONS

1. .375 m
3. 44.8 cm
5. $175.49
7. .875"

9. $86,126.62
11. $55.44
13. $65.87
15. $2,035.65

17. $2,445.38
19. 3"
21. $93.21
23. 54.6875"

UNIT 12 SUBTRACTION OF DECIMAL FRACTIONS

1. $2,188.28
3. $0.55
5. 0.139
7. 1.30 gallons
9. 2.13 feet

11. 2,340.21
13. $728.46
15. $163.65
17. $158.97
19. 33.75 hours

21. 3³⁄₁₆; yes
23. $33.75
25. 37.9 hours
27. 2.5 squares
29. $403.52

UNIT 13 MULTIPLICATION OF DECIMAL FRACTIONS

1. 5.1625
3. .0002178
5. 365

7. 206.25 (207 firebricks)
9. 37.44 hours

11. 9.75 hours
13. 2,106 blocks

UNIT 14 DIVISION OF DECIMAL FRACTIONS

1. 1.233 inches
3. 5.638
5. $27.32
7. 40

9. 50
11. 6.4
13. 64,000

15. 4.5 inches
17. 48 minutes
19. 16 pieces

UNIT 15 EXPRESSING COMMON FRACTIONS AND MIXED NUMBERS AS DECIMALS

1. 0.375
3. 0.25
5. 25/32
7. 17/32
9. 1.25 pounds
11. 0.4375 inch
13. 7/16 inch

15. a. 0.5 inch
 b. 0.042 foot
 c. 0.625 inch
 d. 0.052 foot
 e. 171.554 feet
17. 141.413 feet
19. 146.097 feet

21. 5/8 inch
23. H.I. = 107.55 feet
25. 102.62 feet
27. 3.26 feet
29. 3.10 feet
31. 2.99 feet
33. 2.87 feet

SECTION 4 PERCENT AND PERCENTAGE

UNIT 16 SIMPLE PERCENT AND PERCENTAGE

1. a. 580
 b. 68.75 m
 c. 70%
 d. 91.696
 e. 125%
3. 3,060 square feet
5. 1,350 square feet
7. 119.4 pounds

9. 1,407 bricks
11. 145.6 board feet
13. $41,250
15. $15,000
17. 49.2%
19. a. $17,945
 b. $21,825 material
 c. $8,730

21. 14,350 board feet
23. $7,392.00
25. $2,576.00
27. $3,845.04
29. $243,599.07
31. $86,897.87

UNIT 17 INTEREST

1. $108 interest; $408 amount
3. $128.40 interest; $4,408.40 amount
5. $153.90; $1,233.90
7. $197.14

9. 12.5%
11. $75,250
13. $1,047.17
15. $10,000
17. $1.42
19. a. 4 periods

 b. 2 periods
 c. 0.75% or .0075
 d. $9
 e. $1,209
 f. $9.07
 g. $1,218.07

UNIT 18 DISCOUNTS

1. $4,989.60
3. $45.03
5. 75%
7. $1,250.00
9. $5,640.00

11. $7,000.00
13. $19.83
15. $4.89 discount
 10% rate
17. 3%

19. $631.14
21. $1,204.50
23. a. $1,693.56
 b. $281.44
25. $2,843.62

SECTION 5 MEASUREMENT; DIRECT AND COMPUTED

UNIT 19 LINEAR MEASURE

1. a. $3\frac{1}{8}$"
 b. $2\frac{3}{8}$"
 c. $4\frac{5}{8}$"
 d. $4\frac{1}{8}$"
 e. $3\frac{1}{4}$"
 f. $1\frac{3}{8}$"
 g. $4\frac{1}{2}$"
 h. $3\frac{9}{16}$"
 i. $4\frac{1}{8}$"
3. a. 84 mm
 b. 120 mm
 c. 25 mm

 d. 73 mm
 e. 104 mm
5. 72"
7. 100 m
9. 168 km
11. 9' 8"
13. 58'
15. 50'
17. 740"
19. 86 mm
21. 42'
23. 45.04 m

25. 8' $4\frac{17}{32}$"
27. 40' $10\frac{3}{32}$"
29. 69' $1\frac{12}{32}$" = 69' $\frac{3}{8}$"
31. 53.4072 '
33. 91.106 '
35. 70'
37. 62'
39. 25'
41. a. 22'
 b. 20'
43. 8,046'

UNIT 20 WORKING WITH FEET AND INCHES

1. 44'
3. 165"
5. $237\frac{3}{4}$"
7. 3' 3"
9. 24' 9"

11. 14' $6\frac{5}{8}$"
13. 12' 2"
15. 7' $8\frac{1}{2}$"
17. 4' 5"

19. 3' 7"
21. 12' 11" (spaced at 3
 intervals)
23. 82' 0"

UNIT 21 SQUARE MEASURE

1. 121 square feet
3. 275,625 square feet
5. 35 square feet
7. 39.4662 square feet

9. 238 square feet
11. 814 square feet
13. 6,863.5 square feet
15. 1,338.75 square feet

17. 49.835 square feet
19. 316.561 square feet
21. 2,514.555 square yards

UNIT 22 SURFACE MEASUREMENT—TRIANGLES

1. 176 square inches
3. 250 square feet
5. 638 square feet

7. 203.063 square feet
9. 54 square feet
11. 360 square feet

13. 783.984 square feet
15. a. 275 square feet
 b. 725 square feet

UNIT 23 THE FRAMING SQUARE

1. 9 feet 1⅜ inches

3. 15 feet 7⅞ inches

5. 12 inches rise per foot of run

UNIT 24 SURFACE MEASUREMENT—IRREGULAR FIGUES

1. 59.25 square feet
3. a. 363.8472 square feet

b. 148 square feet
5. 385.5 square feet
7. 1,520 square inches

9. 23.68 square feet
11. 18.713 square feet
13. 38.3 square yards

UNIT 25 SURFACE MEASUREMENT—CIRCLES

1. 7 square feet 10 square inches
3. 50.266 square meters

5. 198 square inches
7. 547.424 square feet
9. 188.692 square feet

11. 9.621 square feet
13. 6.4657 square inches

UNIT 26 VOLUME MEASUREMENT—CUBES AND RECTANGULAR SOLIDS

1. 729 cubic feet
3. 708.9369 cubic feet
5. 120 cubic meters
7. 3 cubic feet

9. 896 cubic feet
11. 3,792.59 cubic yards
13. 1.48 cubic yards

15. 2,870.4 cubic yards
17. 8.796 cubic yards
19. 103.5 cubic feet

UNIT 27 BOARD MEASURE

1. 45 board feet
3. 930 board feet
5. 352 board feet

7. 194.16 board feet
9. 18⅔ board feet

11. 21⅓ board feet
13. 1468.32 board feet

UNIT 28 VOLUME MEASUREMENT—CYLINDERS

1. 12.57 cubic feet
3. 10730.143 cubic feet
5. 62.832 cubic feet

7. 15.29 cubic yards
9. 20.95 cubic feet

11. 25.1436 cubic feet
13. 276.823 cubic inches

UNIT 29 WEIGHT MEASURE

1. 35.25 pounds
3. 1590.9 pounds
5. 20,300 pounds
7. 121.93 pounds

9. 21.33 pounds
11. 240 pounds
13. 129.375 pounds

15. 10.10625 tons
17. 1,209.6 pounds
19. 900 pounds

SECTION 6 POWERS AND ROOTS

UNIT 30 APPLICATIONS OF EXPONENTS IN FORMULAS

1. 64 square inches
3. 49 mm^2
5. 20.97 square feet
7. 2814.87 cubic inches

9. 13.96 cubic feet
11. 256 square feet
13. 144 square feet

15. Foundation: 89.54
 cubic feet
17. 110.61 cubic feet

UNIT 31 USING SQUARE ROOT TO FIND SIDES OF RIGHT TRIANGLES

1. 9
3. 13
5. 23
7. 10' 9⅝" = 10' 9¾"

9. 9' 1⅜" = 9' 1¼"
11. 52' 5¾"
13. 15'
15. 10'

17. 9' 8"
19. 14' 5"
21. 19' 3"

SECTION 7 ESTIMATING

UNIT 32 GIRDERS, SILLS, AND FLOOR JOISTS

1. 9 pieces
3. 150 board feet
5. a. 12 girders

5 b. 2880
7. 32 board feet
9. 288 feet

11. a. 50 joists
 b. 1,333.33 board feet

UNIT 33 ROUGH FLOORING

1. 5 board feet

3. 1,008 square feet

5. 1,890 square feet

UNIT 34 WALL PLATES

1. 36 board feet

3. 36 8-foot lengths,
 12 12-foot lengths

5. 20 lengths

UNIT 35 STUDDING AND FIRE OR DRAFT STOPS

1. 8 linear feet

3. 96 studs

5. 1,056 linear feet

7. 284 studs

UNIT 36 RAFTERS

1. a. 7 feet 10 inches
 b. 352 board feet

3. 68 rafters

5. 48 feet

7. 48 feet

UNIT 37 SHEATHING

1. 270.4 square feet

3. 38 sheets

5. 2,400 square feet

UNIT 38 TRIM

1. 27 feet

3. 36 feet

UNIT 39 ROOFING

1. 35 bundles

3. 352.5 pounds

5. 7.93 square

7. 2 rolls

UNIT 40 DOORS AND WINDOWS

1. $749.76

3. $492.75

5. $155.02

UNIT 41 SIDING

1. 2,656 square feet

3. 2,726 square feet

5. 720 square feet

7. 135 square feet

UNIT 42 STAIRS AND INTERIOR DOOR JAMBS

1. 15 risers

3. 7 inch unit rise,
 10½ inch unit run

5. 16 risers

7. 4½ inches